MAN-MADE
GLOBAL WARMING?

MAN-MADE GLOBAL WARMING?

**IT'S FOOLISHNESS
IN WORDS THAT ALL CAN UNDERSTAND**

Tom Shipley

AuthorHouse™
1663 Liberty Drive
Bloomington, IN 47403
www.authorhouse.com
Phone: 1-800-839-8640

© 2011 by Tom Shipley. All rights reserved.

This book may not be reproduced without the written permission of the copyright holder. However, feel free, with recognition of authors, to use any part of the book to inform the public of the foolishness of man-made global warming.

Published by AuthorHouse   10/12/2012

ISBN: 978-1-4670-3872-0 (sc)
ISBN: 978-1-4634-4273-6 (e)

Library of Congress Control Number: 2011916943

Any people depicted in stock imagery provided by Thinkstock are models, and such images are being used for illustrative purposes only.
Certain stock imagery © Thinkstock.

This book is printed on acid-free paper.

Because of the dynamic nature of the Internet, any web addresses or links contained in this book may have changed since publication and may no longer be valid. The views expressed in this work are solely those of the author and do not necessarily reflect the views of the publisher, and the publisher hereby disclaims any responsibility for them.

Chapter One

Let's Be Objective

I am not a climatic scientist; I am an experienced electrical engineer, drawn to study this subject by the peculiarity of the news I was observing. And it became obvious, very quickly, that something weird was going on. So I began to collect data, news, and other factual information that gradually became available. I have now assembled a wide range of proven, factual details on global warming (which became known as "climate change" when warming stopped in 1998 and temperatures began to drop), the claims and counterclaims that have been published over the years, *and* some of my own observations arising from an engineering point of view. My purpose is to put all the facts I have assembled in one location, for ease of observation and analysis, and to show the absolute foolishness involved.

Others who also became interested have seen much of this information before, reported as individual elements or events and responded to, one at a time. In each case, global warming enthusiasts have responded to each individual weakness reported with the comment that "Yes, that was one little problem . . . but with all things considered, man-made global warming is 'settled science.'" It will be interesting to see their responses to the entire range of factual information, some of which are listed below, as developed by reputable scientists and engineers. All are presented in detail and absolutely refute the idea that "man-made global warming is 'settled science.'"

- Surface and marine temperature measurement procedures totally lack the accuracy necessary to detect the small variations involved. We show that absolutely.

- Global temperatures have not been affected by the amount of carbon dioxide in the atmosphere—they have no relationship to each other. We present data to show that absolutely. Global warming cannot possibly be caused by a man-made increase in CO_2. Humans' activities are exonerated absolutely.
- Satellite measurements have been shown to be highly inaccurate—all records since their installation in 2000 are exposed as highly erroneous. Reports from the Great Lakes area illustrated that.
- The terrible environmental disasters created by global warming, as presented by UN reports, are shown to be untrue.
- The glaciers haven't disappeared, as reported—failing sensors missed an area the size of California.
- The "scientists" improper activities to change temperature-record data to suit their contentions have been exposed; details are shown absolutely. Russia helped with the revelations.
- The global-temperature data from the four responsible agencies do not agree. The agencies do not share expertise, and the differences in their data, one to the other, differ more than the small variations in contention. The average yearly temperature rise from the year 1850 to that of 2008 was only 1.38°F. The existing sensor locations will not allow measurements to that degree of accuracy.
- Temperatures were higher one thousand years ago than they have been in the twentieth century. The efforts of "scientists" are shown as they struggle to lower the estimates of temperatures of those early years.

    We will always have a climate that is changing, warming and cooling, but those changes will not be the result of human activity. You will soon realize why the temperature magnitudes involved were essentially never included in news reports. And the small increments of temperature rise, some hotly disputed, could not even be recognized by the instrumentation, mostly located in unlikely spots, for measuring surface temperatures in the United States. The sensor-reported data, used and manipulated by the four responsible agencies—land based, marine, or satellite—were not sufficiently accurate to dependably measure even the difference between the

Man-Made Global Warming?

coldest year (1911) and the hottest year (1998), which was less than two degrees Fahrenheit during the entire 158-year history of record.

I have not contributed to the science of this information, so I am making every effort to recognize the contribution of those responsible for the data that I am presenting. This will allow you, if questions arise, to check any questionable details. (If you are computer literate, you should know that my greatest accomplice in investigations to obtain this information was Google.) But I have found in past exercises of this type that credits get lost because of the many revisions I have to make for clarification. (I don't write—I rewrite.) If some credits get deleted (if you detect something amiss), just let me know.

Chapter Two

In Selecting a Scientist

There is no recognized selection process currently in use to select scientists. Generally, however, they are people who have gone to school longer than most, and their grades may have been good, indicating intelligence. There are tests to establish the level of intelligence, but the tests are very dependent on memory, and the level of intelligence observed, unfortunately, does not measure how well the mind works—how well it can accept a series of facts, analyze them, and arrive at a correct conclusion.

Scientists flourish in the scholastic field; they go to work for universities to study and analyze subjects that interest them and for which there is a need. Universities do environmental work for the government, and observance of weather conditions—rainfall, temperature, winds, tides—is one of the areas. Times can be tough, economically, for this area of expertise, and universities need government grants to finance their activities. Stimulation is needed, and when times of danger appear, things become more interesting. So, to stimulate an increase in their finances, scientists need to present the public with something in their area of work that threatens the public welfare. They have done that several times in the past and are doing it currently with "climate change." Climate change is their graduation from yesterday's global warming.

In the '20s and '30s, when most people had not received a formal education, those educated people who lacked the ability to properly sort things out and did outrageous things, completely beyond the realm of common sense, were sometimes labeled "educated fools." It is possible that some of our scientists fit this category. But I believe there is another more important reason for choosing an improper course of action.

We Tend to Believe Scientists and Those Labeled "Experts"

In 1998, a British surgeon named Andrew Wakefield published a paper claiming that the measles-mumps-rubella (MMR) vaccine might cause autism. To support his case, Dr. Wakefield reported the stories of eight children who had developed symptoms of autism within one month of receiving MMR. He proposed that measles vaccine virus travels to the intestine, causes intestinal damage, and allows for brain-damaging proteins to enter children's blood streams.

The problem with Dr. Wakefield's study—published in the Lancet, a leading medical journal—was that it didn't actually study the question . . . He wasn't just wrong, he was spectacularly wrong. Moreover, some of the children in his report had developed symptoms of autism before they had received the vaccine—and others never actually had autism.

Dr. Wakefield's paper created a firestorm. Thousands of parents in the United Kingdom and Ireland chose not to vaccinate their children. Hundreds of children were hospitalized . . . and Dr. Wakefield's claim sparked a general distrust of vaccines. In recent years—as more parents chose not to vaccinate their children—epidemics of measles, mumps, bacterial meningitis and whooping cough swept across the United States. The whooping cough epidemic currently raging in California is larger than any since 1955.

In addition, as journalist Brian Deer found, Dr. Wakefield received tens of thousands of pounds [English money] from a personal-injury lawyer in the midst of suing pharmaceutical companies over MMR. (After Mr. Deer's discovery, Dr. Wakefield admitted to receiving the money.) Last year, when the Lancet found out about the money, it retracted his paper. But it was far too late.

Although it's easy to blame Andrew Wakefield, he's not the only one with dirty hands. The editor of the Lancet, Richard Horton, sent Dr. Wakefield's paper to six reviewers, four of whom rejected it. That should have been enough to preclude publication. But Mr. Horton thought the paper was provocative and published it anyway.

Many others in the media showed similar poor judgment, proclaiming Dr. Wakefield's paper an important study even though it was merely a report of eight children that, at best, raised an untested hypothesis.

In the *Wall Street Journal* on January 11, 2011, Paul A. Offit wrote the above (excerpted from his article, *Junk Science Isn't a Victimless Crime)*, which describes the misbehavior of a medical surgeon (scientist). It is a warning to us: all people—including scientists, medical doctors, engineers, etc.—have human weaknesses. We have to know that and to be on guard at all times. Mr. Offit concluded his article with this observation, "The American astronomer and astrophysicist Carl Sagan once wrote that, 'Extraordinary claims should be backed by extraordinary evidence.' Dr. Wakefield made an extraordinary claim backed by scant evidence. Undoubtedly, bad science will continue to be submitted for publication. And if it is provocative enough, some member of the mainstream press, or most of them, will present it to the public."

If your news of world happenings is obtained from the mainstream press, remember: conflict and startling news creates readership. Reporters are taught that in journalism school. Unfortunately, we have to learn the weaknesses of the press for ourselves—with a little help from observers such as Messrs. Offit and Sagan.

What more can you do? Well, there is a book that will help: *Wrong: Why Experts Keep Failing Us—And How to Know When Not to Trust Them,* by David H. Freedman, publisher Little, Brown and Company, Hachette Book Group (Amazon). It covers scientists, finance wizards, doctors, relationship gurus, celebrity CEOs, high-powered consultants, health officials, and more. It won't solve the problems raised by experts in the future, but it will help us recognize some of them. My suspicions are always aroused when any speaker or author in question is labeled "expert."

A Problem

In reports on scientific, engineering, economic, and other technical areas, most media reporters can only report what has been said or written by others. Their education has principally consisted of learning how to write well. During my engineering career with GE, writing about our activities was important, and we were given formal instruction in oral and written presentations. We were taught the skills but were informed that it was more important to know the subject—what we were writing about—and to write clearly, than to know how to write or present a story in the very best way.

Chapter Three

Why You, Individually, Need to Know

Man-made global warming is more than a farce—it is a scam. But it will continue, with billions being spent worldwide, unless enough of you—individual citizens of the United States—pay attention and get actively involved. At this point, too much money is in the air, and it won't disappear until normal political actions are interrupted.

Global warming has occurred. At one time in the distant past, the world was covered with ice, so we know the world has warmed. We know that all warming isn't bad; in fact, past warming has been a blessing for the peoples of most nations. But there is no credible evidence that people's activities have had anything to do with that. During the medieval period—1000AD, for example, before fossil use and other CO_2-generating activities arrived on the scene—global temperatures were higher than they have been since 1900. (Some will disagree with that, but I think the information provided herein will be sufficient to convince you that they are wrong.) Before that, for millions of years, the earth's temperatures cycled—increased and decreased greatly—several times. We will examine these events as we move along and provide a graphic on the final pages that shows temperatures and variations that occurred during the past for millions of years.

In Words That You Can Understand

First, let's examine global warming during our time—the period in question. Are the billions of dollars our politicians are spending to minimize man-made damage to the environment foolish or smart

Man-Made Global Warming?

expenditures? Are the world's glaciers melting and the seas rising at such an alarming rate that steps must be taken now?

The information provided herein will allow people with no engineering or scientific background to decide for themselves. If they can read English, add and subtract, understand small decimal numbers (the latter is necessary because the temperature rise and fall being argued about is so tiny), and understand the markings on indoor or outdoor thermometers, this information will suffice. Together, we will examine the claims of man-made warming, made by a relatively small group of scientists, and the factual findings of a much larger number of scientists that entirely disagree with them. No science or scientific analysis will be required.

Engineers and scientists have their own jargon, and they speak and write of temperature in anomalies, degrees Celsius, Kelvin, and length measurements in meters, centimeters, and millimeters to let their peers know that they belong to the clan. Other nations have departed from their past and have accepted the Celsius scale and metric system as their standards, but the United States has not joined them in the use of these and other systems of measurement. However, I thought you should have a little information on them.

The Fordham Preparatory School's temperature scales website explains the different scales and where they are used. (Google: Fordham Preparatory School's temperature scales for the site.)

The Fahrenheit Scale—The Fahrenheit scale is the scale that is used when they report the weather on the news each night. It is the temperature scale that you are most familiar with, if you live in the United States. The thermometers that you have in your house, for uses such as swimming pools, cooking, bath tubs, or reading body temperature, all incorporate the Fahrenheit scale. In Canada and most other countries, the news will report the temperature using the Celsius scale.

The Celsius Scale—The Celsius scale is commonly used for scientific work. The thermometers that we use in our laboratory are marked with the Celsius scale. The Celsius scale is also called the Centigrade scale because it was designed in such a way that there are 100 units or degrees between the freezing point and boiling point of water. One of the limitations of the Celsius scale is that negative temperatures are very common. Since we know that temperature is a measure of the kinetic energy of molecules, this would almost suggest that it is possible to have less than zero energy. This is why the Kelvin scale was necessary.

The Kelvin Scale—The International System of Measurements uses the Kelvin scale for measuring temperature. This scale makes more sense [than Celsius or Centigrade] in light of the way that temperature is defined. The Kelvin scale is based on the concept of absolute zero, the theoretical temperature at which molecules would have zero kinetic energy. Absolute zero, which is about -273.15°C, is set at zero on the Kelvin scale. This means that there is no temperature lower than zero Kelvin, so there are no negative numbers on the Kelvin scale. For certain laboratory calculations with which we are not concerned, the Kelvin scale must be used.

	Conversion		
Set Points	**Fahrenheit**	**Celsius**	**Kelvin**
water boils	212	100	373
body temperature	98.6	37	310
water freezes	32	0	273
absolute zero	-460	-273	0

The United States still rejects the Celsius scale and retains the Fahrenheit scale for the measurement of temperatures. If the professionals were wise *and wanted you to understand*, they would speak and write using Fahrenheit temperatures—those with which you are familiar. However, it is interesting that we have gotten to this stage of the global-warming phenomenon with essentially no indication of temperature magnitudes—in degrees Celsius or Fahrenheit.

Chapter Four

Look Back a Little

From 1960 to 1975, technology evolved from electron tubes to transistors, to integrated circuits, to microprocessors and computers, and this evolving technology also became evident, later, in advances in radio, TV, cellular telephones, and the personal computer. But it gave humankind the greatest benefits from its use in industry. Using those evolving electronic tools, engineers transformed the way work was being performed. Tedious work, done manually for centuries, was minimized or eliminated. Automated machines and processes, controlled by digital control techniques, replaced laborious manual operations and made work easier.

During my engineering days (1950 to 1965) with General Electric Company, writing about evolving technology and industrial automation activities was extremely important. Those who could use and needed automation didn't know anything about it, machinery equipped with the new systems was much more expensive, and labor unions in the manufacturing sector, quite strong in those days, were adamantly opposed to its use. For these reasons, the company gave engineers, not known to be effective communicators, formal instruction in a wide range of subjects, one of which was effective presentation (oral and written). We were taught writing skills, but the importance of knowing what we were writing or speaking about was emphasized. It was more important to know the subject—what we were writing or speaking about, and to write or speak with clarity—than it was to have journalistic abilities: to be able to present the story in the very best manner.

In those early days, beginning in 1960, the technology evolved much too fast for the professional writers of trade publications to keep up, and the entire industrial arena that they served was thirsting for knowledge on the latest developments. All business and manufacturing procedures previously used in the industrial environment were in the midst of monumental changes.

As the technology evolved, development engineers designed modules and circuitry for use in controls to automate machines and processes, and product engineers modified them to make them suitable for mass production. Application engineers worked with machinery engineers to get an understanding of their machines and how they operated so that the application engineers could then define the motors and controls necessary to motivate and control them properly. They then wrote specifications, including a bill of material, to describe the motors and control functions that had to be provided.

After the equipment described in the specifications was manufactured and shipped, factory engineers supervised the installation, but application engineers were also there. They had the complete story on the machinery; they had defined the control system and described the work the machinery was supposed to perform. They had the responsibility to make sure that the control system and machinery were all in accordance with the specifications.

Electrical engineers did considerable work in mechanics. Prior to the 1940s and '50s, electrical engineering studies in colleges and universities incorporated mathematics that were not required in the mechanical engineering field at that time. The higher mathematics and calculations had evolved to meet the technical demands of electrical power generation, high-voltage transmission systems, and other associated areas. The need for it in mechanical engineering had not yet arrived. Machine automation changed that.

When the automation of machine tools began, it was found that machine tools, satisfactorily designed for manual operation, were not sufficiently tight and rigid to accommodate automatic operations from digitally directed servo systems. It became obvious, shortly, that almost all the machine designs then in use in industry would have to be redesigned if they were to be automated. The machine designers, mechanical engineers mostly, lacked the mathematical skills necessary, so in the '60s, the analyses of existing machine tool construction was performed primarily by electrical engineers. For an application engineer and product planner for GE's Industrial Control

Man-Made Global Warming?

Department, much of the work with machine tool manufacturers in those early days involved working with their machine engineers to help make their machines more responsive to the demands of digital control. One of the services GE offered to them was to analyze their machines' construction—existing ones or others still in the design stage—for rigidity and lost motion. Using their machines' detailed drawings, GE development engineers who specialized in that activity would make the study and analysis, and then meet with the machine design engineers to discuss the findings and proposed changes. It was a very successful service, developed to help advance the market for machine automation and to promote GE's control systems.

Digital control technology was quickly accepted in all sectors of manufacturing, and the United States led the way to the transformation, throughout industry, from manual operation of machines and processes to Numerical Control (NC) of them, using transistor technology to replace relays and electron tubes (1960), to minicomputer directed NC (CNC) of them (1972), to the microcomputer directed NC (CNC) 1975/76 which revolutionized CNC technology. America became the most productive nation on the face of the earth. But America also helped move the rest of the world into this more productive concept. In 1961, for example, GE initiated a program to train foreign engineers of several nations in the benefits and use of the new automated control systems, and in 1969, Monarch Machine Tool Company started work with people in Europe and Japan to train them in the use and benefits of its digitally controlled machines.

Some say that those companies were only interested in others because of greed, but I wouldn't say that. What we presented to the rest of the world was welcomed, and what they gave us in return was also welcomed. That is actually the pleasing phenomenon about capitalism—our free enterprise, free market, natural business system: When a deal is completed, the seller and buyer, both members, mostly walk away satisfied. One leaves with what he got, and the other with what he was paid. The twentieth century oversaw the greatest improvement in quality of life that civilized people have seen in all of history, but neither I nor anyone I knew realized the magnitude of the changes that were taking place.

Tom Shipley

Application engineers of those times knew the total story of the automated machinery and their functions and were there to see the improvements made. They had the best overall picture of what was going on and could tell others. It took me years before I realized that that was the source of my evolving popularity—not my good looks or pleasing way of speaking and writing. We had had a front row seat to industrial changes, and people wanted to know what we saw.

My objective in writing or speaking was generally to describe some automated phenomenon, improvement in a production cycle, or the production of a miraculous product that I had witnessed, and readers of the publications, extremely interested in the advances being made in automation and machining technology, read my stuff and that of others religiously. It was only necessary to write with clarity and to avoid "engineering-speak."

In the '80s and early '90s, small-computer technology was advancing rapidly, and machine manufacturers and users of the machines thirsted for information. I was engaged by a trade publication to inform readers about these advancements. The editor sought me out because of copious material I had written about computers and automated machines—using words and terms that were understandable to interested people who were not engineers or professionals in industry.

During that early period, advertising and professional writers for trade publications were at a severe disadvantage; technology was advancing far too fast for them. But that didn't last; they were concentrating on the technology and its story, and with time, they became proficient. I wrote for seven years; they advanced, the world caught up, and I wasn't needed any more.

Today's reporters for the mainstream press who write or talk about scientific, engineering, or economic matters have the problem that existed during those early days of evolving technology: they do not have the background experience to separate fact from fiction, and they do not have the desire to do so, which those early reporters had. They can't give you a comprehensive story—they can do little more than tell you what they have been told. That is how, as indicated previously, much of the global warming myth got its foothold. Too many of those reporting the news were unable to detect that there were large holes in many of the tales they were being told.

Man-Made Global Warming?

Those of the scientific press, who have a parallel function to that of the early trade publications, have failed in their job. Contrary to the help trade publications of the '40s through the '70s were giving to spread the good word, the scientific press has been busily discouraging the efforts of all who had views, scientifically presented, that differed from the views of their favored few.

After a little investigation into global warming, it became obvious that those writing and speaking about the dangers of global warming belonged in one of two categories: they either didn't know or they had an agenda of their own. Scientists they were not. They, with the popular press writers, collectively labeled the scientists who were pushing the man-made global warming concept as "scientists," and those scientists who were presenting contrary, factual information—even though they too were scientists—as "skeptics" or "deniers."

Chapter Five

Much of the Gloss Is Gone

The mainstream press has done little to present the facts, but the man-made warming premise has lost much of its gloss. The failed December 2009 climate summit in Copenhagen calmed some of the surging political moves toward environmental changes, and many of those with no skin in the game began to pay closer attention. In 2010, those who had begun paying more attention spoke, Congress forced President Obama to back away from major climate legislation in the United States, and French president Nicolas Sarkozy backed away from imposing a climate tax. In Germany, *Der Spiegel* polls have shown a shift from 58 percent (2007) of its citizens who feared climate change to 42 percent (2010), and they appear to be less willing to pay for efforts to protect the environment from warming. The hacked e-mails of scientists in 2009 revealed some extremely damaging activity; too much of the work of the United Nations Intergovernmental Panel on Climate Change (IPCC) unit has been shown to be incorrect and extremely sloppy; and many claims, such as failing glaciers, rising seas, etc., have been shown to have no reputable sources. NASA's data for the United States' climate, which showed the United States to be a major contributor to global temperature increases, were found to be incorrect—the max occurred in 1934—long before the surge of industrial activity began in 1950.

These missteps and others, even with the popular press's seeming lack of attention to them, have resulted in questions about the credibility of some of those engaged in the science of climatology.

Then, when the Senate failed to pass a mandatory federal cap on carbon emissions in July 2010, the Chicago Climate Exchange

(CCX), formed in 2003 as North America's only cap-and-trade system for greenhouse gases, shut down in November 2010. The *Wall Street Journal* (November 20, 2010) pointed out, ". . . The exchange got off to a blazing start with hundreds of companies—from DuPont to Ford to Motorola—voluntarily agreeing to buy and sell rights to emit CO_2 above a legally binding quota. At its peak in 2008, CCX was trading 10 million tons of carbon permits per month, [and] the price of carbon offsets rose from $1.00 per ton to a high of $7.40 in mid-2008. The market collapsed in 2009 when the price fell to $1 . . ." when Congress's lack of interest in cap and trade became apparent."

This was a market the political-powers-that-be had designed. They allotted emission values to all companies. Company A has a top (a cap) to its allowable CO_2 emissions, but it exceeds the cap and can't reduce emissions economically. Company B is another company that has a cap but no problem; it does not need the value allotted to it. Company A can approach Company B and buy the allotment it needs to be okay. The powers that be will, each year, reduce the allotment to all companies, and eventually, all will be in compliance and all global warming will, at that time, have disappeared.

Chapter Six

The Environment Defense Fund explained the problems with CO_2 and the need for cap and trade this way:

> Unlike with some pollutants, all CO_2 goes into the upper atmosphere and has a global—not local—effect. So it doesn't matter whether the factory making the emission cuts is in Boston, Baton Rouge, or Berlin; it reduces global emissions.
>
> Companies can turn pollution cuts into revenue. If a company is able to cut its pollution easily and cheaply, it can end up with extra allowances. It can then sell its extra allowances to other companies. This provides a powerful incentive for creativity, energy conservation, and investment—companies can turn pollution cuts into dollars.
>
> The option to buy allowances gives companies flexibility. On the other hand, some companies might have trouble reducing their emissions, or want to make longer-term investments instead of quick changes. Trading allowances gives these companies another option for how to meet each year's cap.
>
> The same amount of pollution cuts are achieved. While companies may exchange allowances with each other, the total number of allowances remains the same. Nationally, the hard limit on pollution is still met every year.[1]

Can we believe the words of the Environment Defense Fund? I will let you decide. The organization is dependent on funding it

---

[1] "How Cap and Trade Works," last modified November 9, 2010, accessed August 1, 2011, http://www.edf.org/article.cfm?contentID=9112.

Man-Made Global Warming?

can glean from environmental concerns; global warming fears have been good for it.

One fact should have made every journalist who reported on the warming story very suspicious: with all reported increases said to be so dangerous, none included how much of an increase occurred or the actual temperatures involved. For that reason, the public didn't realize how ridiculously little the average global temperatures had changed during the past 158 years. Almost universally, if someone says to you from afar, "Man, it was warm yesterday," you want to know how warm, and I always ask, "What was the temperature?" Why didn't the reporting journalists ask that? I don't know the answer, but that would be a good question to ask your news reporter the next time you see him or her. If the actual temperatures involved had been widely reported by the popular press, in terms that the US public could understand, the foolishness of global warming would never have gotten a foothold.

Lately, however, temperatures have begun to show up in media reports, and you will note there is disagreement among them. There are four authorized agencies that receive grants to work with data, and the data widely presented in this treatise is from one of them—the Climate Research Unit (CRU) of the University of East Anglia, UK. It has published digital tables of monthly and yearly temperature data for public consumption for several years. I found it on the Internet in late 2007.

Chapter Seven

Science and Temperature Data Files

You might need to know some facts about published tables of global warming temperature data—as supplied and used by the scientists. The numbers presented in their tables are not actual temperature magnitudes, and that fact is not revealed in any of the accompanying material. (As of the time of this writing, I have not discovered any public posting of this revelation.) The numbers are labeled "anomalies"; I knew that an anomaly was an irregularity, but that didn't help to alleviate my ignorance. And to further that ignorance, accompanying material included printed charts labeled "Temperature Anomalies (°C)." I worked with that data, thinking that they were absolute temperatures in degrees Celsius, until I noted an Associated Press article on May 18, 2010, that reported that the National Oceanic and Atmospheric Administration (NOAA) had reported that April 2010 was the hottest April in history and, surprisingly, gave the magnitude as 58.1°F. That temperature was too high; according to my data and calculations, average global temperatures were in the realm of 32°F to 33°F. I contacted the Climate Research Unit of the University of East Anglia in Norwich, UK, the source of my data, and Dr. Phil Jones, head of the CRU operation, straightened me out. He explained that to obtain a temperature magnitude from the published CRU numbers, a constant (14) must be added to each of the numbers (anomalies). That sum then becomes the actual absolute temperature in degrees Celsius.

Using Mr. Jones's information, I was able to revise my calculations, check for the hottest year in history, and confirm that the news report was incorrect; the hottest global April temperature in history occurred in 1998–58.284°F (0.602 anomaly, 14.602°C); it was

0.184°F hotter. The NOAA followed the report on April with one on May—it too was hot, but this time, the NOAA acknowledged that May 1998 was hotter.

This example illustrates the extremely small variations that are being argued about in this global warming controversy—variations that are far too small to be measured with the degree of accuracy necessary. The locations of the measuring stations and other variations throughout the many areas of the world prevent the degree of precision that measurement of these small variations would require. And there are variations in precision between the services: the CRU provides data to three decimal points; NASA data that I have seen are presented to only two. The NOAA is quoted frequently by the AP, but I challenge you to find published historical data from the NOAA that is necessary to evaluate their offerings. While you are looking, check on National Climatic Data Center (NCDC) data; I find some curves, but I can find no published tables of digital data.

Most of the raw data are provided to the other agencies by the NCDC—a part of the NOAA known as the Global Historical Climate Network (GHCN). Neither the raw data nor the adjusted data are available to the public; only after the responsible agencies have adjusted the data and the station anomalies have been averaged for 5x5 grids is the public invited in. (We don't need to know what a grid is, but let me tell you what the NCDC says it is: Each month of data consists of 2,592 gridded data points produced on a 5×5 degree basis for the entire globe [72 longitude by 36 latitude grid boxes].)

The Associated Press story concerning the hot April in 2010 was misinformation, widely distributed by the popular press. Most of the nation's popular news outlets obtain information on national and foreign affairs from the AP. In this case, as in many others, instead of checking the validity of the NOAA information using Internet CRU data available to everyone, the AP merely passed on the information it received to all the newspapers that subscribed to its services. And the frequency with which it has presented these warming stories in the absence of truth makes me wonder: Is it possible the AP doesn't know?

Chapter Eight

The Temperatures

The maximum global temperature rise, from the very coldest year to the very warmest during the past 158 years of record, has been somewhere around 1.98°F. Most of the thermometers we own are graduated in two-degree increments, and we can't read the temperature changes within that range with them. The lowest and highest temperature magnitudes reported during the twentieth century vary between CRU- and NCDC-reported data, but the difference (the rise) and the years in which they occurred remain about the same. The variation in temperature magnitudes results from differences in calculations used by the four designated agencies.

The Goddard Institute for Space Studies (GISS) is a third group with designated responsibilities. It reported its record of global data on December 16, 2008, and it showed wide variations in magnitude and years from CRU and NCDC data. The lowest temperature in GISS data occurred in 1918, and its magnitude was -0.4 anomaly, 13.6°C, 56.48°F; the highest was for the year 2005 with a magnitude of 0.62 anomaly, 14.62°C, 58.32°F; and the total rise was 1.84°F.

". . . The analysis of the individual temperature-sensor measurements that are reported by the NCDC from all parts of the world is difficult," the agencies say, "because the measuring instrumentation is not evenly distributed throughout the world, measuring technology has changed over the years, the environment in which some measurements are made (rural to city) have affected measurements, and other reasons."

Information will be provided to you later that chides the scientists for the slipshod approach to sensor location, which severely affects measuring accuracy."

The Earth Warms Up

The 1.98°F average global temperature rise is calculated from digital data provided by the CRU. It shows the lowest average global temperature on record occurred in 1911 and the temperature was -0.573°C anomaly, 13.427°C, and 56.169°F. That is the coldest year on record during the 158 years of measurement history. The highest temperature on record occurred in 1998, and it was 0.529°C anomaly, 14.529°C, and 58.152°F. (The CRU curves in figure 1 do not agree with its tabulated digital data, and there is no explanation. The temperatures differ widely. The lowest temperature in the curve is -0.47°C anomaly while the digital data report shows -0.57; the highest temperature in the curve is 0.5°C anomaly, though the digital data show 0.526.) The digital data say the highest temperature rise was 1.98°F; the curves say the rise was 1.75°F. The NCDC shows the same years with the highest and lowest temperatures, 58.1 and 56.336 respectively, and a temperature rise of 1.98°F also (fig. 1).[2]

A rise during a different period of time, which would be more useful in my opinion, could be found by examining the average global temperature for the very first year, 1850, and that of the very last, 2008, to determine the difference—how much the temperature had risen. That is actually just as interesting and pertinent as the lowest and highest, since neither actually determines a trend. This approach shows the temperatures to have been -0.439 anomaly, 13.561°C, and 56.41°F in 1850, and 0.327 anomaly, 14.327°C, and 57.79°F in 2008 for a rise of only 1.38°F.

---

[2] The curves are presented online by the NOAA and the NCDC. Is it possible that the author accidentally labeled them incorrectly? NOAA Satellite and Information Service, "The Instrumental Record of Past Global Temperatures," last updated October 4, 2010, accessed August 1, 2011, http://www.ncdc.noaa.gov/paleo/globalwarming/instrumental.html.

Chapter Nine

Pick a Period—and Predict Terrible Things

There are other approaches to defining temperature rise. During the recorded history of temperature over the 158-year period, various scientists selected particular periods that seemed to foretell a phenomenon in which they were interested—and which later proved to be incorrect. In the 1970s, for example, scientists picked their period and predicted calamitous cooling. George Will told us a little about it with the article "The Law of Doomsaying," February 15, 2009, Washington Post.

". . . In the 1970s, 'a major cooling of the planet' was 'widely considered inevitable' because it was 'well established' that the Northern Hemisphere's climate 'has been getting cooler since about 1950' (the *New York Times*, May 21, 1975). Although some disputed that the 'cooling trend' could result in 'a return to another ice age' (the *Times*, September 14, 1975), others anticipated 'a full-blown 10,000-year ice age' involving 'extensive Northern Hemisphere glaciation' (*Science News*, March 1, 1975, and *Science* magazine, December 10, 1976, respectively). The 'continued rapid cooling of the Earth' (*Global Ecology*, 1971) meant that 'a new ice age must now stand alongside nuclear war as a likely source of wholesale death and misery" (*International Wildlife*, July 1975). 'The world's climatologists are agreed' that we must 'prepare for the next ice age' (*Science Digest*, February 1973). Because of 'ominous signs' that 'the Earth's climate seems to be cooling down,' meteorologists were 'almost unanimous' that 'the trend will reduce agricultural productivity for the rest of the century,' perhaps triggering catastrophic famines (*Newsweek* cover story, *The Cooling World*, April 28, 1975). Armadillos were fleeing south from Nebraska, heat-seeking snails were retreating from central European forests, the North Atlantic was 'cooling down about as fast as an ocean can cool,' glaciers had 'begun to advance' and 'growing seasons in England and Scandinavia are getting shorter' (*Christian Science Monitor*, August 27, 1974) . . ."

The extreme temperatures during 1998 (warmest) and 1911 (coldest) are demonstrations by Old Mother Nature, necessary from time to time, to tone us down when we begin to exhibit our weakness for human power. El Niño, responsible scientists tell us, was responsible for the warmth upheaval in 1998–and the trade winds that caused the warmth reversed course and gave us global coolness for several months thereafter.

Chapter Ten

The World Is Coming to an End

Political leaders met in Kyoto, Japan, in December of 1997 to consider a world treaty to restrict human production of "greenhouse gasses," primarily CO_2. They said CO_2 would create "human-caused global warming" if allowed to increase further, with disastrous environmental consequences. By 2009, most nations had signed on to the Kyoto Protocol, which meant that they agreed with the "consensus" that the world's use of energy was creating levels of CO_2 that were a danger and must be curtailed.

The Environmental Defense Fund (EDF), on April 19, 2010, (http://apps.edf.org/page.cfm?tagid=54192) formally introduced this proclamation concerning global warming:

> Despite overwhelming scientific evidence, popular myths and misinformation abound. Here are the facts of what we know about global warming.
>
> FACT—There is scientific consensus on the basic facts of global warming.
>
> The most respected scientific bodies have stated unequivocally that global warming is occurring, and people are causing it.
>
> FACT—The global warming we are experiencing is not natural. People are causing it.
>
> Only CO2 and other greenhouse gas emissions from human activities explain the observed warming now taking place on Earth.

Man-Made Global Warming?

FACT—Glaciers are melting and are a contributor to sea-level rise.

Between 1961 and 1997, the world's glaciers lost 890 cubic miles of ice. The consensus among scientists is that rising air temperatures are the most important factor behind the retreat of glaciers on a global scale over long time periods.

FACT—Global warming and increased CO2 will harm many economies and communities.

While some skeptics may argue that there are benefits to global warming and extra CO2, warming in just the middle range of scientific projections would have devastating impacts on many sectors of the economy.

FACT—Many communities won't be able to adapt to rapid climate change.

The current warming of our climate will bring major hardships and economic dislocations—untold human suffering, especially for our children and grandchildren.

For more information, see our in-depth scientific report [PDF] on the myths and facts of global warming by Dr. James Wang and Dr. Michael Oppenheimer.

Posted: 19-May-2010; Updated: 19-Apr-2010The most respected scientific bodies have stated unequivocally that global warming is occurring, and people are causing it.

Chapter Eleven

Let's Look at NCDC and CRU Temperature Charts

The first chart of global temperatures, figure 1 below, presents NCDC data as a plotted curve. The second chart, marked CRU, is produced by the Climatic Research Unit. Note that in the year 1998, NCDC shows a temperature that reaches about 0.57°C (58.226°F), while the CRU shows it to reach about 0.50°C (58.1°F).

The CRU curve shows temperatures increasing after 1998. This is in defiance of all data published on this subject, and the digital data from the same organization show a decided decline in global temperatures since the high of 1998.

Scary Spikes

The temperature curves shown look scary; observe the spikes, the large up and down swings. The curves are scary, no question about it. They show a very rapid rise—very steep—during past years, but you should know that the curves do not accurately reflect temperatures as we know them.

Man-Made Global Warming?

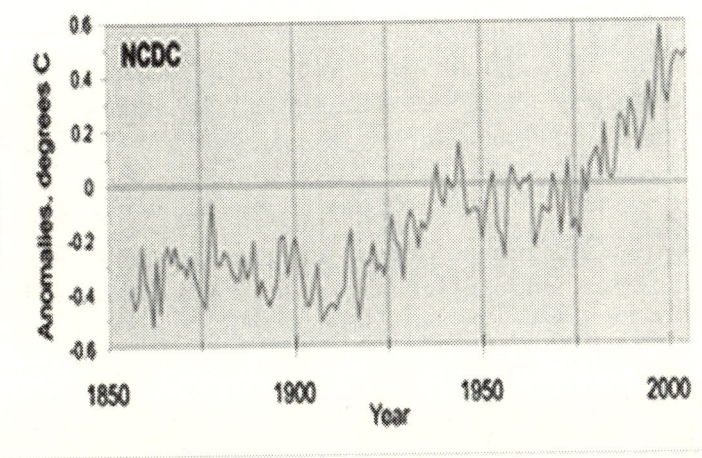

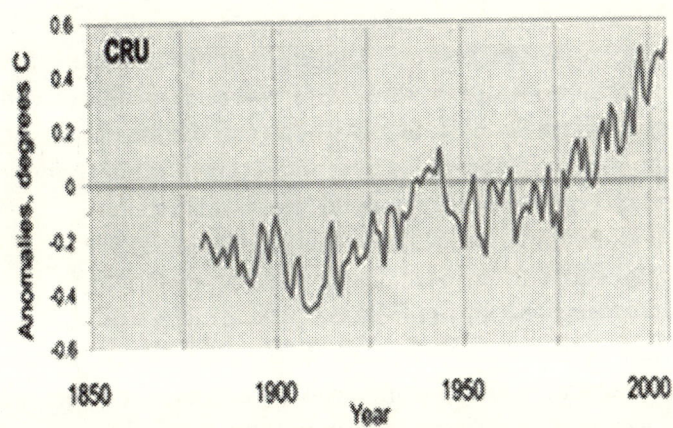

Fig 1. NCDC and CRU Global Temperature Anomalies by Year

The scientists are plotting anomaly data. When the anomaly doubles (0.2 to 0.4, for example), the temperature appears to double. But when 14 is added to the anomalies to make them absolute temperatures in degrees Celsius, the change from 14.2 to 14.4 looks far less dangerous when plotted. When converted to degrees Fahrenheit—to terms with which the people of the United States are familiar—the fear should go away completely.

The total temperature rise, as mentioned before, has been approximately 1.98°F. Think about that: All of that bristling activity,

Tom Shipley

during 158 years or so, record temperature variations that change within a tiny two-degree Fahrenheit variation. As an engineer, I examine the scary nature of these graphs and it occurs to me that for once, our politicians may have done something right; they refused to change from our old tried-and-true measurement methods while the rest of the world has selected poorer standards. Ounces, inches, and degrees Fahrenheit make more sense than grams, millimeters, and degrees Celsius. Scientists may have a need, in their work, for extremely small increments (1 inch equals 25.4 millimeters, 1 ounce equals 28.35 grams, 1°C equals 33.8°F, etc.), but in my opinion, those increments are too small for reasonable, everyday use for measurements.

And in view of the small magnitude of change for 158 years, think about the accuracy of next week's weather prediction; I believe that you will realize that neither scientists nor any other human being can predict whether average global temperatures ten, twenty, or one hundred years out will be either warmer or colder within two degrees Fahrenheit. Ridiculous. There are too many uncertainties, variables, and self-interested scientists and organizations involved who depend on the public dole for their livelihood.

Global Temperatures—161 Years

Most readers, for the first time, will see actual global temperatures, as they bounced and changed, during their yearly time of record. The three charts of temperature add information to the two scary charts shown in figure 1, which the climate scientists call *Anomalies*.

These records, covering 161 years of actual, measured global temperatures, show the increase has been insufficient to cause worries regardless of whether man-made or not.

I began work to convert the Anomaly records accumulated during 2009, 2010, and 2011 to Celsius and Fahrenheit scales. The Anomaly data were published by the Climate Research Unit of the University of East Anglia (UEA), Norwich, UK, one of the four agencies responsible for climate studies, and I had recorded them in Excel computer spreadsheet files.

Man-Made Global Warming?

On my second day of plotting, my son T. Doane Shipley, an engineer with Eaton Corporation in Pittsburgh, interrupted my activities with a telephone call, and when I told him what I was doing, he said, "Dad, that can be done easily with Excel." He then told me how I could do it. I have used Excel for years, but I had a new computer, new operating system, and a new Excel program, and I couldn't get it done. We talked some more, and I finally sent him the data by e-mail. He did the work, and the fruits of his labor are shown in Figure 2.

You will find, as you read the book, that Doane's work is another example of the methods I used to write it. I mined other people's work, dependable sources for specific, factual information, to illustrate the problems and to show that we are spending billions of dollars, along with the rest of the world, on a fool's quest. The foolishness was foisted on the popular media by so-called scientists, probably in error at first, but then continued in efforts to increase funding for the work they do. The popular press then foisted it on the public – world-wide. We should blame the scientists for their misdeeds, but shouldn't forget the Associated Press – its news reports, without any searching questions or knowledgeable investigation, spread the foolishness and made it real.

The top graphic on page 32 is a plot of *anomalies* that the global-warming scientists have published widely. An anomaly is actually a temperature measured in degrees Celsius, from which the number 14 has been subtracted. This, not explained in any writings I have seen; it is a fact which I obtained from Dr. Phil Jones, University of East Anglia, Climate Research Unit (CRU).

(It must be understood why the dates of published data are referred to often in the text. The reason is this: The data change from date to date. No great changes, but changes. I don't know why the anomalies for 1850 are still changing, but the charts say they are. Charts dated 2/26/2009, used when I started the Book, showed -0.439 Anomaly for 1850; charts dated 8/12/2012 show -0.435. The scientists are playing with small numbers, so we have to pay attention.)

Tom Shipley

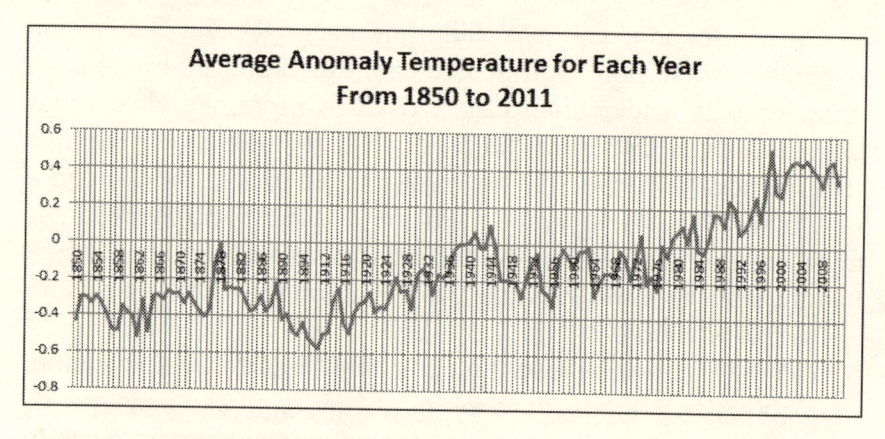

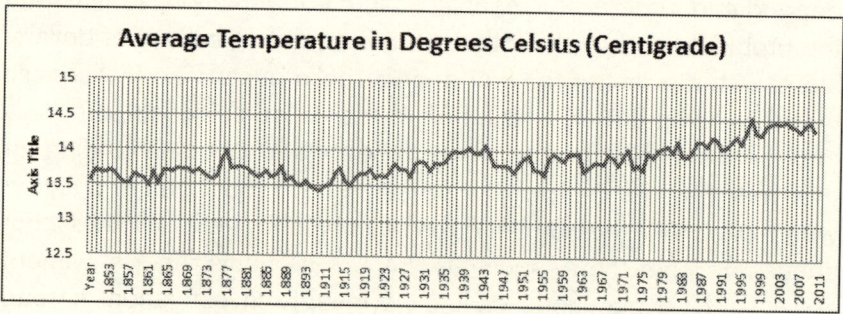

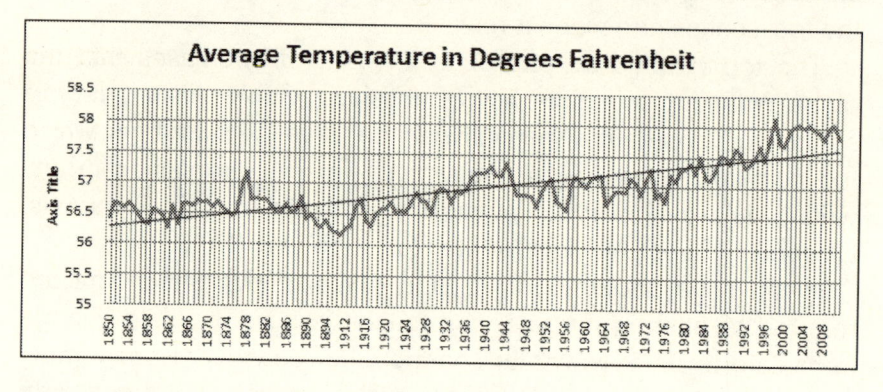

Figure 2

An Anomaly plot has to show positive and negative readings, so a reference point, 0, is necessary, and in this case is somewhat unusual—0 is actually a temperature of 14 Deg. Celsius (equivalent to 57.2 Deg. F in the United States). The middle graph shows the plot that results from converting anomalies to absolute temperature

Man-Made Global Warming?

in degrees Celsius—by adding 14 to each anomaly. You can see how the conversion to degrees Celsius smooths the curve—takes away some of its threatening qualities. The Celsius scale, known as the Centigrade scale until 1954, is used in United States laboratory environments and in all areas of the rest of the world.

The above plot of Anomalies exhibits an increase or decrease far more definitively than a plot of actual temperature measurements. For example, in 1982 the anomaly was 0.015 and the temperature increased until 1998, to an anomaly of 0.529. Plotting this shows an increase of 34.27, or 3,427%; But reference to the middle graph, the Celsius scale, shows that the actual Celsius temperature for those years went from 14.015 Degrees C to 14.529, for a change of only .03667, or 3.67%, and the lowest graph, the Fahrenheit scale, the temperature increased from 57.23 Degrees F to 58.15—only 0.01607—a 1.61% increase in actual temperature.

You cannot work to that degree of accuracy by referring to the charts. But I can; I used the actual data for that period. (However I should note: The actual temperatures with which I am working, which the scientists have provided, were not measured with sufficient accuracy. So I'm not working to the necessary degree of accuracy, either. The measurement sensor problems are explained later.)

The bottom graph shows the average yearly global temperatures plotted in degrees Fahrenheit. It is difficult to show graphically, but that introduces a further, but much smaller, smoothing effect from that shown in the middle Celsius graph.

A previous scientists' fear; the earth was cooling dangerously, was discussed earlier. In the 1970s, "a major cooling of the planet" was "widely considered inevitable" because it was "well established" that the Northern Hemisphere's climate "has been getting cooler since about 1950"

Checking the top chart during this time, it shows that Anomalies had bobbled up and down as they dropped during that period, but had decreased from 1944 until 1976. By the time the scientists had noted the downward trend, determined that a danger existed and were really into publicizing it—it was gone. Anomalies increased in 1977, continued to increase until 1998, and then began to drop. But they remained considerably above the scientists' dangerous 1944 to 1976's anomalies for 35 years.

Tom Shipley

The Anomalies graph shows a series of temperature cycles, beginning in 1850, when temperatures were first admitted to the record. From 1850 to 1878 temperatures increased; from 1878 to 1911 they dropped; from 1911 to 1944 they increased; from 1944 to 1976 they trended down; from 1976 to 1998 they increased; from 1998 to 2011 they have decreased—from 0.529 to 0.348. But a reference to the CO_2 chart on page 38, reviewing these yearly periods of global-temperature rise and fall, it is evident that CO_2 and temperature are ignoring each other. The chart on page 37, in CO_2 in parts per million shows the same. Real Science is correct: *". . . We can state with 100% certainty that as CO_2 increases, temperatures will either go up, go down, or stay the same."*

With Microsoft's Excel program, a trend line can be shown. The program averages all the temperature ups and downs and provides a linear line to show the average of the increases and decreases of all the plotted data from the beginning to the end. It shows the total increase, from 1850 to 2011, to have been 1.4 degrees Fahrenheit. (Interestingly, the rise from the first year, 1850, to the last year, 2011, was very close to that: 1.399 degrees F.) The book shows you that the location of the sensors in the United States cannot provide measurements to that degree of accuracy.

Chapter Twelve

Humankind's Energy Use and Increased Levels of CO_2

Now, let's look at the known levels of CO_2. First, I should tell you how accurate the data is that is shown in the two graphs in figure 3. ZFacts (http://zfacts.com/p/194.html) tells us that:

> Considering the difficulties of gathering data from centuries past, the data is amazingly close. We know, for example, that up until 1950, deforestation was putting about as much CO_2 into the atmosphere as humans were while burning fossils, but it's not easy to know how many trees were chopped down in, say, 1850, and how much CO_2 this put into the air. In spite of imperfect data the fact that a simple calculation predicts the extraordinary shape of actual CO_2 so well is clear evidence this can not be a coincidental convergence. Human CO_2 emissions must have caused the upsurge in atmospheric CO_2.
>
> Where did the figure 3 numbers come from? CO_2 levels. These come from two sources. From 1958 forward, they are from a weather station high atop the Mona Loa volcano in Hawaii. They are so accurate, they show levels going up every autumn, when the leaves fall, and coming down every spring. Earlier data are from ice cores in Antarctica. The two sources agree remarkably well. (All actual data points are shown in the top graph.) The next graph shows the emissions data.
>
> Fossil fuel data are from the US Department of Energy, including an estimate that fossil fuel emissions will increase at 1.2 percent per year until 2030. We assume this rate to 2050. However, the average rate for the last five years was about 1.9 percent, so this may be optimistic. Cement manufacture

also releases CO_2 from the rocks it processes, so that source is tracked separately.

Deforestation data extended only from 1850 to 2000, so zFacts fitted a straight line to the first one hundred years (1850-1950) to extend the data back to 1750. Deforestation peaked in 1991 and from 1992 to 2000 followed a steep and straight descent. Again, a straight line was fitted to predict the future. As can be seen above, this is a dramatic and possibly optimistic prediction.

CO_2 Emissions and Global Warming

In spite of the reservations expressed by the much larger number of the world's scientists, world politicians posing as scientists maintain that temperatures are rising, and the increases are directly related to the amount of CO_2 generated by people. The energy used by the world's people to keep warm, travel, and work is causing the problem.

Let's examine that theory—using the data we have right here. Curves of CO_2 emissions from 1750 to 2000 and global temperatures from 1850 to about the same time are shown. If the "scientists" are correct, a comparison of the curves of NCDC's and CRU's global temperature with that for CO_2 (provided by zFact.com) should show that the trend of temperature parallels that of CO_2. If CO_2 emissions are increasing, then temperatures will also be increasing; and if the temperature decreases, it will be due to a drop in CO_2 levels.

It does not require a scientist's advanced education to see that CO_2 increased rather sharply from about 1880 to 1937 and then skyrocketed in 1950. But average temperatures decreased during that time until about 1920. CO_2 paid no attention whatsoever; it increased consistently from 1 billion metric tons of carbon to about 1.7 billion metric tons. Now, look at the curves once more. The average temperature from 1940 to 1975 decreased and then increased to remain relatively constant, but the amount of CO_2 rose tremendously; it rose from 2 billion metric tons to about 6 billion metric tons. The magnitude of CO_2 emissions obviously failed to

Man-Made Global Warming?

drive global temperatures in the pattern described by the global warming enthusiasts.

Australian Senator Steve Fielding, in his attempt to be reelected in 2010, used the work of Dr. Milo Wolff and Dr. Richard Malotky to show the record of global CO_2 magnitudes and temperatures for the years 1995 to 2009. The good doctors used CRU temperature data, the Mauna Loa Observatory for CO_2, and produced a composite chart. It showed CO_2 rising steadily form 365 parts per million in 1998 to 385 ppm in 2008, while the temperature in 1998, 58.152°F, decreased to 57.79°F in 2008. That small drop in temperature of 0.362°F is very small, but it represents a drop of 19% of the maximum rise in temperature in 158 years, and a 26% drop for the beginning and end of the 158-year record. And once again; CO_2 and temperature ignored each other.

These scientists' views should be discredited even more than they have been at this point. The only reason the views haven't been labeled criminal is that our politicians have passed no laws to make them so.

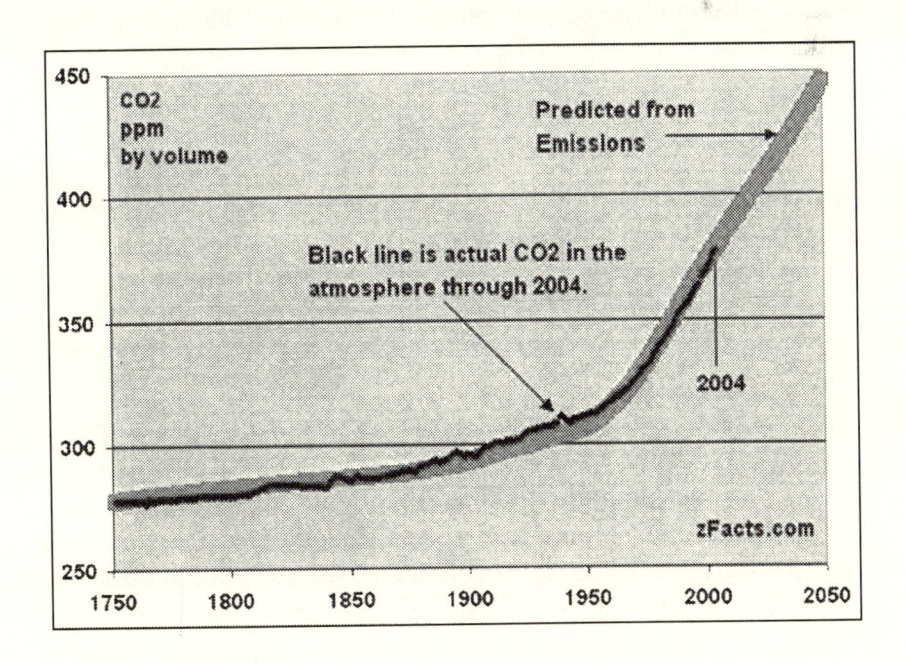

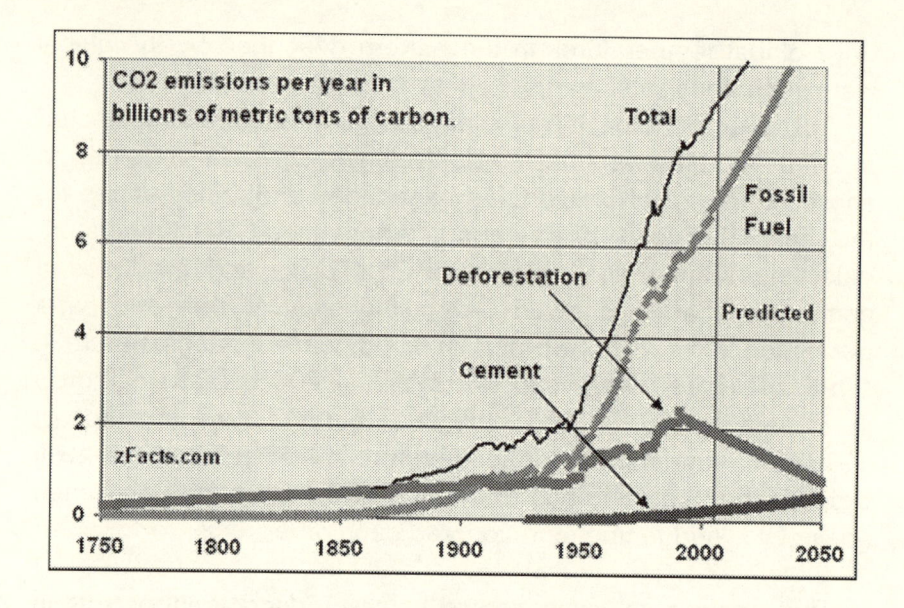

Fig. 3. Source: zFacts.com

In their article "Carbon Dioxide and Global Warming: Where We Stand on the Issue," C. D. Idso and K. E. Idso of the Center for the Study of Carbon Dioxide and Global Change have this to say about carbon dioxide (CO_2) and global temperature changes: (http://www.co2science.org/about/position/globalwarming.php)

> The observation that two things have risen together for a period of time says nothing about one trend being the cause of the other. To establish a causal relationship it must be demonstrated that the presumed cause *precedes* the presumed effect. Furthermore, this relationship should be demonstrable over several cycles of increases and decreases in both parameters. And even when these criteria *are* met, as in the case of solar/climate relationships, many people are unwilling to acknowledge that variations in the presumed cause truly produced the observed analogous variations in the presumed effect.
> In thus considering the seven greatest temperature transitions of the past half-million years—three glacial terminations and four glacial inceptions—we note that increases

Man-Made Global Warming?

and decreases in atmospheric CO_2 concentration not only did not precede the changes in air temperature, they *followed* them, and by *hundreds to thousands of years*! There were also long periods of time when atmospheric CO_2 remained unchanged, while air temperature dropped, as well as times when the air's CO_2 content dropped, while air temperature remained unchanged or actually *rose*. Hence, the climate history of the past half-million years provides absolutely no evidence to suggest that the ongoing rise in the air's CO_2 concentration will lead to significant global warming.

Now, let's go back 600 million years and see what was happening.

Real-Science, at the site, http://www.real-science.com/correlating-co2-temperature-geologic-record, shows a plot of CO2 (in atmospheric parts per million) and temperature (in degrees Celsius) for 600 million years. It shows that for 600 million years of our history, CO2 and earth temperatures have had absolutely no relationship to each other.

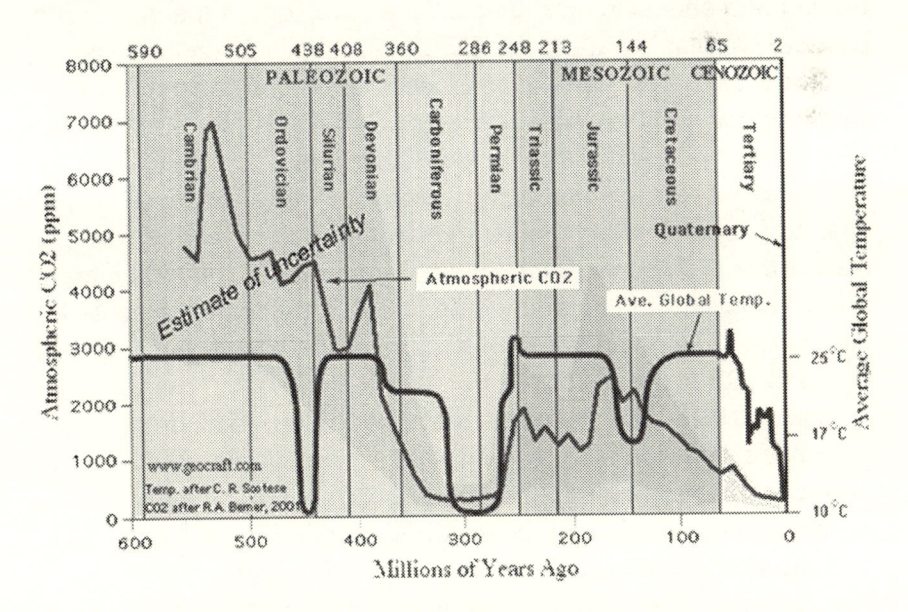

Figure 4 (10°C = 50°F; 25°C = 77°F)

Tom Shipley

The authors say, "During the Ordovician [period] CO2 was more than ten times higher than at present. Global temperatures ranged between very hot and an ice age. We can state with 100% certainty that as CO2 increases, temperatures will either go up, go down, or stay the same."

Our warming-enthusiast scientists' warnings of world devastation from man-made global warming were not scientifically based and showed a dangerous ignorance of world history. And our popular press spread the word. Now, just as they shifted from global-warming to climate-change when it was evident that temperatures had dropped since 1998, you are beginning to see, as this information is made public, a shift from CO2 as the global-warming culprit, to methane or some other gas as the cause.

The information is available from other sources. If you have a compelling interest in the history of the earth, this site and www.scotese.com/climate.htm are Internet sites in which to spend some time. Don't pay any attention to the usual suspects—NOAA, NASA, etc.; they still need funding to satisfy their lifestyle, and since their past work has been shown to be disreputable, they have been picking and choosing. For temperature data, ignore theirs; use the publically available data at the CRU Internet site given in the book.

Some die-hards have questioned the plotted temperatures in Fig. 4 and where they came from. Mr. Scotese explains at his site, and some of his work is presented in the book.

Others have questioned the validity of CO2, because of the methods used to reveal magnitudes for millions of years. For them, we offer more recent information. Temperatures, as Anomalies, degrees Celsius and Fahrenheit since 1998, the hottest globally for the entire global-temperature history, is provided in Fig. 11, pages 107 and 108 in the book.

Since 1998, temperatures have wobbled, but dropped in magnitude. If the two waltz together, as alarmists claim, CO2 should have followed suit and trended down also. David Evans' work shows, definitely, there is no togetherness. His temperature and CO2 data, starting in 2002, show that even though CO2 is rising definitively, temperatures have trended down during the entire period.

Man-Made Global Warming?

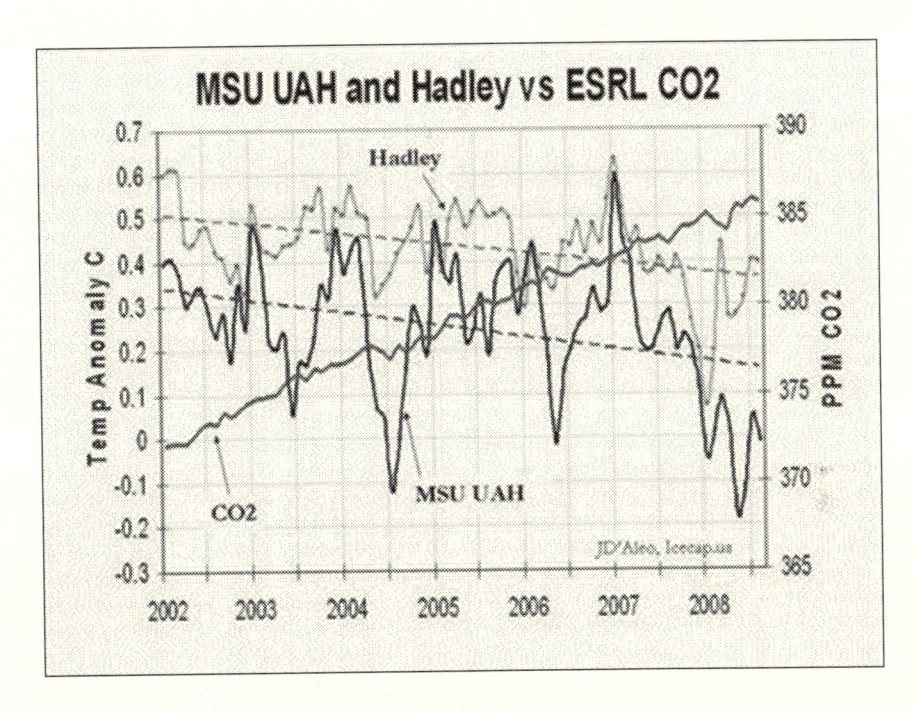

MSU UAH and Hadley vs ESRL CO2

Figure 5

Mr. Evans shows that there is no correlation between temperature, CO2, or fossil fuel use. For more information: http://diehardempiricist.blogspot.com/2012/04/global-warming-is-bunk-part-5-no.html

The temperatures shown do not agree with the CRU data that I have been observing since 2009. However, using my data for 1998 'til 2011 (as posted 8/12/2012), the temperature anomalies varied from 0.529 Anomaly in 1998 to 0.339 in 2011 (58.155°F to 57.810). A drop of 0.345°F. Small—but that represents a drop of 24.76% of the 1.393°F increase that had occurred from the year 1850 to the year 2011 or 17.41% of the 1.98 maximum temperature change (lowest to highest) during those years. It is obvious that temperatures dropped during the thirteen year period, yet CO2 went sharply in the opposite direction, as you can see from the above chart. You can also see the sharp increase in CO2 during this period from the charts on page 33 and 34. For more information on CO2 and global-temperature correlation, visit: http://cosmoscon.

Tom Shipley

com/2012/02/28/global-temperature-and-co2-update-march-2012.

It should now be apparent; as the Real-Science authors said: "We can state with 100% certainty that as CO2 increases, temperatures will either go up, go down, or stay the same."

Chapter Thirteen

The Scientific Consensus: No Man-Made Warming

This climate theory, construed and backed up by the IPCC (and some number of scientists), has been disavowed by a far greater number of other scientists. Official news at the Global Warming Petition Project tells us:

> 31,487 American scientists have signed this petition, including 9,029 with PhDs . . .

> We urge the United States government to reject the global warming agreement that was written in Kyoto, Japan in December, 1997, and any other similar proposals. The proposed limits on greenhouse gases would harm the environment, hinder the advance of science and technology, and damage the health and welfare of mankind. There is no convincing scientific evidence that human release of carbon dioxide, methane, or other greenhouse gasses is causing or will, in the foreseeable future, cause catastrophic heating of the Earth's atmosphere and disruption of the Earth's climate. Moreover, there is substantial scientific evidence that increases in atmospheric carbon dioxide produce many beneficial effects upon the natural plant and animal environments of the Earth.[3]

This should be sufficient evidence for politicians, if they are honorable, that a serious case has been made: man-made global warming is not "settled science."

---

[3] "Global Warming Petition Project," accessed August 1, 2011, http://petitionproject.com/.

Tom Shipley

The Scientists, Instrumentation, and Measurements?

To measure the tiny incremental changes in temperature involved in all the discussions of global warming that are scaring our well-educated scientists, the instrumentation must be state of the art. The "terribly" dangerous rise in temperatures from 1850 to 2008 has been only 1.38°F. The errors made by NASA that moved the hottest year in the United States from 1998 to 1934 was 59.432°F and 59.450°F respectively—only 0.018°F difference. To detect temperatures of this magnitude will require our measuring sensors to be extremely accurate and their location sites to be selected with extreme care. If you have an outdoor thermometer, you know the difficulty of locating it to prevent direct sunlight from giving erroneous high-temperature results. And you know that it must be located away from an air conditioner exhaust. Are the sensors sufficiently accurate for the purpose, and have the locations been selected with the proper care?

Anthony Watts has made a study of this, he formed surfacestations.org for the purpose in 2007, and his report in 2009 answers those questions. I hate to be the bearer of bad news, but here goes. Mr. Watts begins,

> Global warming is one of the most serious issues of our times. Some experts claim the rise in temperature during the past century was "unprecedented" and is proof that immediate action to reduce human greenhouse gas emissions must begin. Other experts say the warming was very modest and the case for action has yet to be made.
>
> The reliability of data used to document temperature trends is of great importance in this debate. We can't know for sure if global warming is a problem if we can't trust the data. The official record of temperatures in the continental United States comes from a network of 1,221 climate-monitoring stations overseen by the National Weather Service, a department of the National Oceanic and Atmospheric Administration (NOAA).

Man-Made Global Warming?

Until now, no one had ever conducted a comprehensive review of the quality of the measurement environment of those stations.

During the past few years I recruited a team of more than 650 volunteers to visually inspect and photographically document more than 860 of these temperature stations. We were shocked by what we found.

We found stations located next to the exhaust fans of air conditioning units, surrounded by asphalt parking lots and roads, on blistering-hot rooftops, and near sidewalks and buildings that absorb and radiate heat. We found 68 stations located at wastewater treatment plants, where the process of waste digestion causes temperatures to be higher than in surrounding areas.

In fact, we found that 89 percent of the stations—nearly 9 of every 10—fail to meet the National Weather Service's own siting requirements that stations must be 30 meters (about 100 feet) or more away from an artificial heating or radiating/reflecting heat source.

In other words, 9 of every 10 stations are likely reporting higher or rising temperatures because they are badly sited.

It gets worse. We observed that changes in the technology of temperature stations over time also has caused them to report a false warming trend. We found major gaps in the data record that were filled in with data from nearby sites, a practice that propagates and compounds errors. We found that adjustments to the data by both NOAA and another government agency, NASA, cause recent temperatures to look even higher.

The conclusion is inescapable: The U.S. temperature record is unreliable. The errors in the record exceed by a wide margin the purported rise in temperature of 0.7° C (about 1.2° F) during the twentieth century. Consequently, this record should not be cited as evidence of any trend in temperature that may have occurred across the U.S. during the past century. Since the U.S. record is thought to be "the best in the world," it follows that the global database is likely similarly compromised and unreliable.[4]

This report presents actual photos of more than one hundred temperature stations in the United States, many of them demonstrating vividly the siting issues we found to be rampant in the network. Photographs with explanation are provided to show the problems, rampant, that eliminate all possibility of accurate temperature readings.

---

[4] Anthony Watts, *Is the U.S. Surface Temperature Record Reliable?* (Chicago: The Heartland Institute, 2009), http://wattsupwiththat. files.wordpress.com/2009/05/surfacestationsreport_spring09. pdf. Additional information is available at Anthony Watts, "Surface Stations," accessed August 1, 2011, http://www.surfacestations.org.

Chapter Fourteen

Satellite Measurements—Better Data?

But all is not lost. In 2000, our scientists initiated a satellite system for temperature measurements. Measurements from high above the earth should eliminate the problems just described, which are rampant in earth-surface site locations. That was my hope, so I looked for data and information on how measurements are made and how accurate they are. John O'Sullivan, with his Wall Street Journal, August 11, 2010, article, "Satellite Data Failure Means Decade of Global Warming Data Doubtful," ruined my day. Can't this bunch of people get anything right?

>US Government admits satellite temperature readings "degraded." All data taken off line in shock move. Global warming temperatures may be 10 to 15 degrees [Fahrenheit] too high. The fault was first detected after a tip-off from an anonymous member of the public to the climate skeptic blog, Climate Change Fraud (August 9, 2010).
>
>...Caught in the center of the controversy is the beleaguered taxpayer funded National Oceanic and Atmospheric Administration (NOAA). NOAA's Program Coordinator, Chuck Pistis has now confirmed that the fast spreading story on the respected climate skeptic blog is true.
>
>However, NOAA spokesman, Program Coordinator, Chuck Pistis declined to state how long the fault might have gone undetected. Nor would the shaken spokesman engage in speculation as to the damage done to the credibility of a decade's worth of temperature readings taken from the problematic "NOAA-16" satellite.

Tom Shipley

"NOAA-16" was launched in September 2000, and is currently operational, in a sun-synchronous orbit, 849 km (527.5 miles) above the Earth, orbiting every 102 minutes providing an automated data feed of surface temperatures which are fed into climate computer models.

NOAA has reported a succession of record warm temperatures in recent years based on such satellite readings, but these may now all be undermined. World-renowned Canadian climatologist, Dr. Timothy Ball, after casting his expert eye over the shocking findings concluded, "At best the entire incident indicates gross incompetence; at worst, it indicates a deliberate attempt to create a temperature record that suits the political message of the day."

The Great Lakes See Weird, Wild Temperature Fluctuations

Great Lakes users of the satellite service were the first to blow the whistle on the wildly distorted readings that showed a multitude of impossibly high temperatures. NOAA admits that the machine-generated readings are not continuously monitored so that absurdly high false temperatures could have become hidden amidst the bulk of automated readings.

In one example swiftly taken down by NOAA after my [Sullivan's] first article, readings for June and July 2010 for Lake Michigan showed crazy temperatures off the scale ranging in the low to mid hundreds—with some parts of the Wisconsin area apparently reaching 612° F. (Can you imagine that—212 °F is boiling?) With an increasing number of further errors now coming to light, the discredited NOAA removed the entire set from public view. But just removing them from sight is not the same as addressing the implications of this gross statistical debacle.

Man-Made Global Warming?

NOAA Whitewash Fails in One Day

NOAA's Chuck Pistis went into whitewash mode on first hearing the story about the worst affected location, Egg Harbor, set by his instruments onto fast boil. On Tuesday morning Pistis loftily declared, "I looked in the archives and I find no image with that time stamp. Also we don't typically post completely cloudy images at all, let alone with temperatures. This image appears to be manufactured for someone's entertainment."

But later that day Chuck and his calamitous colleagues, now with egg on their faces, threw in the towel and owned up to the almighty gaffe. Pistis conceded, "I just relooked at the image again and it is in my archive. I do not know why the temperatures were so inaccurate. It appears to have been a malfunction in the satellite. We have posted thousands of images since the inauguration of our Coastwatch service in 1994. I have never seen one like this."

But the spokesman for the Michigan Sea Grant Extension, a 'Coastwatch' partner with NOAA screening the offending data, then confessed that its hastily hidden web pages had, indeed, showed dozens of temperature recordings three or four times higher than seasonal norms. NOAA declined to make any comment as to whether such a glitch could have ramped up the averages for the entire northeastern United States by an average of 10-15 degrees Fahrenheit by going undetected over a longer time scale.

Somewhat more contritely, NOAA's Pistis later went into damage limitation mode to offer his excuses.

"We need to do a better job screening what is placed in the archive or posted. Coastwatch is completely automated so you can see how something like this could slip through."

In his statement Pistis agreed NOAA's satellite readings were "degraded" and the administration will have to "look more into this." Indeed, visitors to the Michigan Sea Grant site now see the following official message:

"NOTICE: Due to degradation of a satellite sensor used by this mapping product, some images have exhibited extreme high and low surface temperatures. Please disregard these images as anomalies. Future images will not include data from the degraded satellite and images caused by the faulty satellite sensor will be/have been removed from the image archive."

Blame the Clouds, Not Us Says NOAA

NOAA further explained that cloud cover could affect the satellite data making the readings prone to error. But Pistis failed to explain how much cloud is significant or at what point the readings become unusable for climatic modeling purposes.

As one disgruntled observer noted, "What about hazy days? What about days with light cloud cover? What about days with partial cloud cover? Even on hot clear days, evaporation leads to a substantial amount of water vapor in the atmosphere, particularly above a body of water. How can this satellite data be even slightly useful if it cannot 'see' through clouds?"

Top Climatologist Condemns Lack of Due Diligence

The serious implications of these findings was not lost on Dr. Ball who responded that such government numbers with unusually high or low ranges have been exploited for political purposes and are already in the record and have been used in stories across the mainstream media, which is a widely recognized goal.

The climatologist who advises the military on climate matters lamented such faulty data sets, "invariably remain unadjusted. The failure to provide evidence of how often cloud top temperatures 'very nearly' are the same as the water temperatures, is unacceptable. If the accuracy of the data is questionable it should not be used. I would suggest it is rare, given my knowledge of inversions, especially over water."

Man-Made Global Warming?

How Many Other Weather Satellites Are Also "Degraded"?

A key issue the government administration declined to address was how many other satellites may also be degrading. "NOAA-16" is not an old satellite—so why does it take a member of the public to uncover such gross failings?

Climate professor, Tim Ball, pointed out that he's seen these systemic failures before and warns that the public should not expect to see any retraction or an end to the doom-saying climate forecasts: "When McIntyre caught Hansen and NASA GISS with the wrong data in the US I never saw any adjustments to the world data that changes to the US record would create. The US record dominates the record, especially of the critical middle latitudes, and to change it so that it goes from having nine of the warmest years in the 1990s to four of them being in the 1930s, is a very significant change and must influence global averages."

Each day that passes sees fresh discoveries of gross errors and omissions. One astute commenter on www.climatechangefraud.com noted, "It is generally understood that water heats up more slowly than land, and cools off more slowly. However, within the NOAA numbers we have identified at least two sets of data that run contrary to this known physical effect."

The canny commenter added, "Two data points in question are at Charlevoix, where the temperature is listed at 43.5 degrees—while temperature nearby (+/−30 miles) is 59.2 degrees; and in the bay on the east side of the peninsula from Leland is listed at 37.2 degrees. These are supposedly taken at 18:38 EDT, 6:38PM (19:38 Central, or 7:38PM). These are both taken in areas that appear to be breaks in the cloud cover.

With NOAA's failure to make further concise public statements on this sensational story it is left to public speculation and "citizen scientists" to ascertain whether ten years or more of temperature data sets from satellites such as NOAA-16 are unreliable and worthless.

Chapter Fifteen

What About the UN IPCC's Report

Now that we know the responsible agencies can't even measure temperature with sufficient accuracy to even know what global temperatures are doing, let's look at what has happened since the IPCC released its 1997 report with all its warnings and stated dangers.

The Environmental Defense Fund people responsible for the April 19, 2010, remarks previously quoted had put their faith in the UN's IPCC reports. But they had obviously missed previous news reports that serious flaws had been found in IPCC's allegations.

The IPCC and those responsible for NASA GISS dataset, the NCDC GHCN dataset, the CRU dataset, and the Japan Meteorological Agency dataset, at that time, had to have been in deep disarray. During January and February of 2010, the *Wall Street Journal*, the *Guardian*, the *Telegraph*, and others of the press presented articles that should have concerned the Environmental Defense Fund and others of the global warming activist community. The articles revealed radical errors in the claims made in the UN's IPCC 2007 report. The gist of them were:

Himalayan glaciers. They will not be melting soon; the IPPC claim was withdrawn with apologies. Indian geologist Vijay Kumar Raina, before the IPCC report was published, said, in his report written for the Indian government, that glacier activity was "nothing out of the ordinary." And there were no suggestions that glaciers would disappear. In 2006, months before the paper was published, Georg Kaser, an expert in tropical glaciology at the University of Innsbruck, Austria, and a lead author for the IPCC, warned that the 2035 prediction was clearly wrong. "This [date] is not just a little bit wrong, but far out of any order of magnitude," he said. Rajendra Pachauri, the chair of IPCC, who had no expertise of any kind in climatology, rejected both without investigation. He dismissed the Raina report as not being peer reviewed and said, "With the greatest respect this guy retired years ago and I find it totally baffling that he comes out and throws out everything that has been established years ago." (The material published was not peer reviewed—it was merely a rewrite of a media interview with a scientist who audibly expressed his views on the subject.)

The UN's climate science body has admitted that the 2007 report was incorrect.

But Jean-Pascal van Ypersele, vice chair of IPCC, said that this mistake "did nothing to undermine the large body of evidence" showing human activity a problem of global warming. As we progress and reveal other suspicious problems in the work produced by this great political body supposedly representing science, I think you will disagree with the vice chair's comment. (www.guardian.co.uk/environment/2010/jan/20/ipcc-himalayan-glaciers-mistake/print)

Forty percent of the Amazon forests were endangered. Not so; that was a mistake. The IPCC report said, "Up to 40% of the Amazonian forest could react drastically to even a slight reduction in precipitation . . . the tropical vegetation, hydrology and climate system in South America could change rapidly to another steady state." This, too, was a mistake. Jonathan Leake of London's *Sunday Times* reported that the claim came not from a peer-reviewed science-oriented publication as claimed, but from a report from the World Wildlife Fund (WWF). And the report from the WWF had misrepresented a study from the journal *Nature*, and that study wasn't assessing rainfall—it was looking at the impact on forests from human activity such as logging and burning. (*Wall Street Journal*, Review & Outlook, February 16, 2010)

The UN's climate science body has admitted that the 2007 report was incorrect.

Tom Shipley

Melting mountain ice in the Alps, Andes, and Africa. This was not global warming. The *Telegraph* revealed this on January 30, 2010. These IPCC claims were not made by informed, peer-reviewed scientists, and they were selected by IPCC warming enthusiasts from articles in nonscientific magazines that would advance the cause for global warming. The sources used in this case were a feature article in a popular magazine for mountain climbers, *Climbing*, which merely described the changes they were noticing on the mountainsides around them. The other great scientific source was the thesis written by a geography student studying for a master's degree—he quoted interviews he had with mountain guides.

It appears that the WWF had been a favored source for the IPPC. Information was drawn from it for claims in several other areas, and "It can [now] be revealed that the [2007] IPCC report made use of 16 non-peer-reviewed World Wildlife Fund reports." A few of them posited these problems were caused by man-made global warming.

The Wall Street Journal, Review & Outlook, February 16, 2010, reported these damaging facts:

The transformation of natural coastal areas. Coastal areas are okay; the report was a mistake.

The destruction of mangroves. This is not going to happen—a mistake.

Glacial lake outbursts—mudflows and avalanches. Nope. More mistakes.

Changes in the ecosystem of the Mesoamerican reef. No science was involved—a mistake. These drastic changes, caused by human activity, were all inspired by the WWF, a green lobby for environmental causes. And, as the *Journal* says, "It believes in global warming, and its 'research' reflects its advocacy, not the scientific method." Can we agree there is no peer review here?

Man-Made Global Warming?

Water resources for 4.5 billion people could be depleted by 2085. The study that reported that also informed readers that global warming could increase supplies for as many as 6 billion people. This was actually good news, but it didn't fit the IPPC's purpose, so that detail was not reported.

Fifty-five percent of Holland below sea level. Holland was miffed—only 26 percent was below sea level. The IPCC used Holland as an example in order to advance its cause for stopping global warming. The study stated, "The Netherlands is an example of a country highly susceptible to both sea level rise and river flooding because 55 percent of its territory is below sea level."

Extreme weather-related events, rapidly rising costs. Forget it. A single study, not peer reviewed, was just a minor IPCC mistake. The *Journal* pointed out, "In fact, nobody can reliably establish a quantifiable connection between global warming and increased disaster-related costs."

The danger of continued global warming. Not so fast. Phil Jones, former head of U of E. Anglia (CRU) has informed the BBC "there has been no 'statistically significant' warming over the past 15 years, though he considers this to be temporary." CRU global temperature data, mentioned previously, posted by the university, show a drop since 1998 by 18 to 20 percent of the maximum increase during the past 158 years. Professor Jones is the first weather scientist to make a warming statement that agrees with the global data I have been following since early 2009.

An "indisputable" fact that the 1900s and 2000s were the hottest in 1000 years. Nope. The same *Wall Street Journal* report said the well-known climatologist, Phil Jones, former head of U of E. Anglia (CRU), now says the world may have been warmer during medieval times than now.

The North Pole ice-free during the 2008 melt season. There was lots of coverage, but it was a huge mistake. The sensors used to measure temperature had been failing; that had been causing errors. The *Baltimore Examiner,* November 25, 2009, pointed out that "most coverage of this report fails to state that just this winter the National Snow and Ice Data Center had to come clean that its remote sensors were failing. It had UNDERESTIMATED ice about the size of California." That was the explanation.

1990 China study. The Guardian, February 1, 2010, informed us that weather stations in China were moved substantially but not reported. China and its stations were the basis of a study, important in climate-change science, which purported to show that the effect on temperature by big city urban locations was negligible. In response to allegations of fraud, crucial data from the study cannot be reviewed for verification; the documentation no longer exists. Phil Jones, CRU, says the stations were probably moved, but records underpinning the 1990 study have been lost.

Chapter Sixteen

The World Gets Colder

In fact, during the last ten years, 1998 to 2008, while the "scientists" have been arguing about warming, global temperatures have been hopping up and down but decreasing from the maximum. In 1998, the temperature was 14.526°C, 58.15°F; in 2008, the temperature was 14.327°C, 57.79°F. The drop, 0.36°F, has eliminated 18.2 to 20 percent of the 1.98-degree rise that occurred in 158 years. However, to describe all that happened, year by year: from the maximum in 1998, the temperature dropped to 57.699 in 2000, wobbled back up to 58.055 in 2005, and then decreased in value in 2008. This information is not shown in the curves, but you can see the change, year by year, in the digital temperature data presented in figure 11 below. These tiny changes that we are talking about here can't be detected with the instrumentation presently in place in the United States.

Temperatures Before 1850

I was reading a book, *Kingsport Heritage, the Early Years, 1700 to 1900*, by Murial M. C. Spoden, The Overmountain Press, Johnson City, Tn., 1991, during an October 2009, visit to my hometown, Kingsport, in northeast Tennessee. During an unexpected extension of my stay at Fox Manor, I ran out of reading material, and the proprietors, Susan and Walter Halliday, interested in the history of the area, had a copy of the book and let me read it. They also gave me a list of other books on the city's history.

Tom Shipley

The book introduced me to a most strange phenomenon about temperature changes involving my hometown area: It described a year without summer. About that, the author said,

> Boating was sharply curtailed in the spring of 1816 when Christianville citizens experienced the severe weather called, "the year without a summer." [Christianville, an early name for west Kingsport, was located along the South Fork Holston River, now known as the Old Kingsport area.] The cold weather began in April, and the snow, ice and frost became progressively worse through June, July and August. The cause was traced to a huge volcanic eruption on April 15 on the East Indies Island of Tambora. Dust and debris blew seven to fifty cubic miles into the atmosphere causing a red glow at sunset around the world.
>
> People froze to death [in the Kingsport area] as snow and sleet fell for seventeen days in May and June. August was worse as *ice* coating killed everything green. Ice, half an inch thick, covered ponds and rivers. All crops were ruined. In the spring of 1817, the seed of corn that had been stored in 1815, sold for the very high price of $5 to $10 a bushel.

That did not describe the years of spring and summer as I remembered them during my first twenty-four years of life there, and I wasn't about to believe it without verification. So I went to the Internet and Googled "1816, the year without summer" and found lots of stuff on the event. I settled on: *1816–The Year Without Summer* by Lee Foster, Meteorologist. The whole nation and the rest of the world had had the problem, but there were no definitive records—that occurred before we began with records. One article, which explained the cause of the phenomenon, was written by Foster and was good, but I was particularly attracted to his information on what others of those early days thought had caused the weather change: "Some believed it was caused by sinners, while some blamed it on Benjamin Franklin's lightning rod experiments."

Man-Made Global Warming?

The book doesn't say, but I wondered, could those latter beliefs have been introduced by scientists of those times?

When you realize that a volcanic eruption in such a far-away place can cause such a change in the weather of lands thousands of miles distant, you begin to have a feel for Mother Nature's strength and how little our activities are even noticed in the global scheme of things. Some folks, who spent too much of their time in schools, don't like to feel that unimportant.

Chapter Seventeen

CRU's Record of Global Temperatures

To check the temperature data presented here, or anywhere else, I go to the CRU's website.[5] This is a properly recognized service, there has been no charge, and I haven't had to join any organization. The data are presented in anomaly degrees Celsius. To get to data that you understand, you will have to convert anomaly data to degrees Celsius (°C) and then to degrees Fahrenheit (°F). First, add 14 to the anomaly decimal numbers presented; that gives you °C. Then convert Celsius data to Fahrenheit. To avoid the difficult equation, (DegF=1.8xDegC+32), Google "Fahrenheit to Celsius converter." Insert the Celsius number, click outside the box, and read.

The National Climatic Data Center (NCDC), at (www.ncdc.noaa.gov/paleo/globalwarming/instrumental), says this about the CRU data included here:

> The earliest records of temperature measured by thermometers are from Western Europe beginning in the late 17th and early 18th centuries. The network of temperature collection stations increased over time and by the early 20th century, records were being collected in almost all regions, except for Polar Regions where collections began in the 1940s and 1950s.
>
> A set of temperature records from over 7,000 stations around the world has been compiled by the NOAA National Climate Data Center to create the Global Historical Climatology Network (GHCN) (GHCN Version 2 data set;

---

[5] "Temperature Data (HadCRUT3 and CRUTEM3)," accessed August 1, 2011, www.cru.uea.ac.uk/cru/data/temperature/hadcrut3vgl.txt.

Peterson and Vose 1997). About 1,000 of these records extend back into the 19th century.

The data from the above site are developed by the CRU at the School of Environmental Sciences, University of East Anglia, Norwich, United Kingdom. The CRU provides data in digital form from 1850 to the current year for both land and sea.

A comparable digital data set for sea ice and sea surface only shows the coldest global temperature to have been –13.385°C (56.093°F) in 1910 and the hottest temperature to have been 14.451°C (58.012°F) in 1998–an increase of only 1.919°F.[6] And it too shows a temperature drop since 1998 of 0.36°F–a reversal of 18.75 percent from the total increase of 1.919°F seen in 158 years.

Problems with Media Information

Much temperature data being presented by the popular media is incorrect; an example was given previously. Another set of data in a newspaper report, purported to rebut data presented herein, was shown to be incorrect by the actual table heading presented in the report *Contiguous 48 U.S. Surface Air Temperature Anomaly*. It looks good, demonstrates the author's contention, but is not global; it is surface air data only, and for just the United States.

We do not get much useful information on subjects involving technology from most media sources, but sometimes we go a little too far in criticizing them. Much of the time the failure has been unintentional; the reporters merely reported what they were told. This has been a particular problem with their reports on technical matters involving science–and it has kept us from knowing the actual state of the world's climate.

As indicated previously, in reporting on scientific, engineering, economic, and other technical areas, most media reporters can only relay what they have been told by their sources. Their education has

---

[6] "Temperature Data (HadCRUT3 and CRUTEM3)," accessed August 1, 2011, www.cru.uea.ac.uk/cru/data/temperature/hadsst2gl.txt.

consisted, principally, of learning the language and how to speak and write it well.

NASA's Record of Global Temperatures

NASA has received some notoriety because of missteps, so we should examine its contribution to global temperature analyses as it existed on November 13, 2010.[7] The table is labeled Global Land-Ocean Temperature Index," and from all available information, it should be directly comparable to the CRU's table of global Land-Ocean data.

NASA's work does not agree with the CRU's. In fact, all the digital data produced by the different services responsible for climate information and available for review disagree in too many areas. The following data involving numbers demonstrate that.

The first data listed are for 1880, defined by two digits only (CRU data are to three), and all numbers in the table must be divided by 100 to obtain anomalies comparable to the CRU's. As mentioned previously, to obtain absolute temperatures in degrees Celsius, add 14; and to obtain degrees Fahrenheit, multiply the degrees Celsius by 1.8 and add 32.

In looking for the coldest temperatures, we find -0.39 anomaly, 13.61°C, 56.498°F (a much warmer temperature than that found in the CRU's data), and it occurred for three different years: 1890, 1907, and 1917. The warmest temperature is 0.63, 14.63°C, 58.334°F in 2005, for a maximum rise of 1.836°F. This rise is in the same range as the CRU's data, but NASA's maximum and minimum temperature magnitudes vary greatly from those presented by the CRU, and they occur during entirely different years. The temperature increase of 1.314°F, from 1880 (56.678°F) to 2008 (57.992°F) is in the range of the CRU's for 158 years, but the temperature magnitudes vary greatly.

You are probably beginning to understand why temperature magnitudes have seldom appeared in news reports involving global warming. The different services responsible for monitoring temperature, from the variations, appear to be working from entirely

---
[7]

Man-Made Global Warming?

different sensor data—and perhaps different world locations. The only consensus that I have detected in this affair, from scientists who have made their selves known, is that 1998 was the warmest year on record. Yet NASA's digital data, downloaded from the above site on November 13, 2010, show the warmest year on record occurred in 2005, and at 58.334°F, it was warmer than that shown by the CRU's warmest year of 1998, which was 58.15°F. The difference in the temperature magnitudes shown by these numbers may appear to be very small—and they are—as pointed out previously. The scientists have established the parameters; these small changes appear to have been important in their prediction that the world is heading toward terrible consequences if we don't change our way of living.

Chapter Eighteen

Climate Scientists Hide Procedures

The scientists for all four designated operations do not share the procedures they use in processing raw data to reach their individual conclusions on temperature magnitudes for their published data. I think that since the taxpayers of the United States are funding these operations, we should demand that their procedures be made public. This would allow engineers, scientists, and other members of our citizenry to review the processes and propose even better ones for more dependability. It is obvious those currently involved need help.

Mr. McIntyre, who detected errors in NASA's work for the United States and then the hockey stick problem in global calculations, asked for guidance in the techniques and processes used in both cases. His request was rejected by both groups, and he had to reverse-engineer the details. He convinced NASA about the US problem and revisions were quietly made. The hockey stick foolishness is still maintained. If that is proven to be erroneous, the whole buildup for global warming (climate change, whatever) is down the drain, and funding will suffer dramatically.

United States Industry: Caused Global Warming

For years, NASA and other scientists said the industrial boom in the United States was the driving force in man-made global warming, shown by the fact that our hottest temperature, 59.432°F, occurred in 1998—the same year, as determined by NASA, that the world's hottest temperature, 58.147°F, had occurred. The prolific production activity in the US accompanied by massive amounts of

CO_2, beginning in about 1950, was deemed to be the driving force for it all.

Observe the plotted curves (figure 1); note how the temperature flattens out from 1950 to 1975. 1950 was the beginning of the US industrial boom years. If humankind was contributing greatly to the warmth, that part of the curve should not be so flat; that is when we really began to throw CO_2 into the atmosphere. In 1963, Congress began a campaign to improve our air quality and increased its reach in 1970. Yet the steepest trend in increased temperature begins at about 1975 and continues until 1998. The charts end at 2004 or 2005, but the digital data mentioned above show data on through 2008.

Then, in August of 2007, the United States was exonerated; it was no longer the bad boy in global warming. NASA, the responsible US agency, was forced by the work of Canadian Steve McIntyre to revise the temperature data. The hottest average *global* temperature for a year occurred in 1998, but the hottest *US* temperature, revised from 59.432 to 59.45°F, did not occur in 1998, as previously proclaimed by NASA; it occurred in 1934. (http://www.norcalblogs.com/watts/2007/08/1998_no_longer_the_hottest_year.html)

Look at the miniscule movements of temperature that were in contention. They could not possibly have been measured accurately by the system of sensors and their locations installed in the US and elsewhere. This degree of accuracy was not possible, and had you tried to monitor the average temperature differences, the effort would have been a total failure, of course, and you would have missed both the fight and the celebration. Your thermometers could not have detected it, and your system, your body, could not have felt any difference at all.

Mr. McIntyre also pointed out that four of the top ten hottest years occurred during the '30s, while only three (1998, 2006, and 1999) retained their hot luster within the past ten years. The temperatures of 2000, 2002, 2003, and 2004 disappeared from contention. This comprehensively rebutted scientists' claims that US industrial activity was the moving force for global warming during the last 60 years.

Tom Shipley

The Old US Temperatures In order of hotness:			The New, Revised Temperatures In order of hotness:		
Year	Anomaly	Deg F	Year	Anomaly	Deg F
1998	1.24	59.432	1934	1.25	59.45
1934	1.23	59.414	1998	1.23	59.414
2006	1.23	59.414	1921	1.15	59.216
1921	1.12	59.216	2006	1.13	59.414
1931	1.08	59.144	1931	1.08	59.144
1999	0.94	58.892	1999	0.93	58.874
1953	0.91	58.838	1953	0.90	58.82
2001	0.90	58.820	1990	0.87	58.766
1990	0.88	58.784	1938	0.86	58.748
1938	0.85	58.73	1939	0.85	58.73

New NASA data can be reviewed at (http://data.giss.nasa.gov/gistemp/graphs/Fig.D.txt) and you will note that the data do not agree with that shown above; the anomalies shown for 1934 and 1998 (June 20, 2011) are 1.195 instead of 1.25, and 1.318 instead of 1.23. respectively. No explanation is given for the extreme deviation from the widely published, and argued about, earlier reports. (Data obtained 11/8/2010 were to two decimal points; that obtained 8/23/2011 was to four–but the data, except for very early years, had not changed in magnitude.)

The corrected data should have made the global warming enthusiasts a little more timid, but it didn't, as you can see from the NASA data. However, McIntyre's revelations were useful in demonstrating to those with open minds that something fishy was going on. For the others, it should have been a heavy warning to them to remove the world's industrial activities as a cause of global warming. The only clue to the possibility that the others got the message is the fact that the previous global warming activists have deserted that term and are now worrying about "global climate change."

The error in NASA's records was detected by Steve McIntyre of Toronto, Canada, and he, with difficulty, finally convinced NASA that its data, beginning in 2000, was incorrect. As a result of this revelation and after the revision was made, Dr. John Theon, the supervisor of James Hansen, the scientist responsible for the "error," said in January 2008, "I appreciate the opportunity to add my name to those who disagree that global warming is man-made." (Hansen was one of Al Gore's principal allies in the all-out promotion of man-made global warming.) The news was provided by Watts Up With That on the Internet, and complete details from John S. Theon's e-mails to Marc Morano@EPW.Senate.Gov were posted on the US Senate Committee Internet site. They follow in reverse order:

Sent: Thursday, January 15, 2009 10:05 PM
To: Morano, Marc (EPW)
Subject: Climate models are useless

Marc, First, I sent several e-mails to you with an error in the address and they have been returned to me. So I'm resending them in one combined e-mail.

Yes, one could say that I was, in effect, Hansen's supervisor because I had to justify his funding, allocate his resources, and evaluate his results. I did not have the authority to give him his annual performance evaluation. He was never muzzled even though he violated NASA's official agency position on climate forecasting (i.e., we did not know enough to forecast climate change or mankind's effect on it). He thus embarrassed NASA by coming out with his claims of global warming in 1988 in his testimony before Congress.

My own belief concerning anthropogenic climate change is that the models do not realistically simulate the climate system because there are many very important sub-grid scale processes that the models either replicate poorly or completely omit. Furthermore, some scientists have manipulated the observed data to justify their model results. In doing so, they neither explain what they have modified in the observations, nor explain how they did it. They have resisted making their work transparent so that it can be replicated independently by other scientists. This is clearly contrary to how science should be done. Thus there is no rational justification for using climate model forecasts to determine public policy.

With best wishes, John

Sent: Tuesday, January 13, 2009 12:50 PM
To: Morano, Marc (EPW)
Subject: Re: Nice seeing you

   Marc, Indeed, it was a pleasure to see you again. I appreciate the opportunity to add my name to those who disagree that Global Warming is man made. A brief bio follows. Use as much or as little of it as you wish.
   John S. Theon Education: B.S. Aero. Engr. (1953-57); Aerodynamicist, Douglas Aircraft Co. (1957-58); As USAF Reserve Officer (1958-60),B.S. Meteorology (1959); Served as Weather Officer 1959-60; M.S, Meteorology (1960-62); NASA Research Scientist, Goddard Space Flight Ctr. (1962-74); Head Meteorology Branch, GSFC (1974-76); Asst. Chief, Lab. for Atmos. Sciences, GSFC (1977-78); Program Scientist, NASA Global Weather Research Program, NASA Hq. (1978-82); Chief, Atmospheric Dynamics & Radiation Branch NASA Hq., (1982-91); Ph.D., Engr. Science & Mech.: course of study and dissertation in atmos. science (1983-85); Chief, Atmospheric Dynamics, Radiation, & Hydrology Branch, NASA Hq. (1991-93); Chief, Climate Processes Research Program, NASA Hq. (1993-94); Senior Scientist, Mission to Planet Earth Office, NASA Hq. (1994-95); Science Consultant, Institute for Global Environmental Strategies (1995-99); Science Consultant Orbital Sciences Corp. (1996-97) and NASA Jet Propulsion Lab., (1997-99).
   As Chief of several NASA Hq. Programs (1982-94), an SES position, I was responsible for all weather and climate research in the entire agency, including the research work by James Hansen, Roy Spencer, Joanne Simpson, and several hundred other scientists at NASA field centers, in academia, and in the

private sector who worked on climate research. This required a thorough understanding of the state of the science. I have kept up with climate science since retiring by reading books and journal articles. I hope that this is helpful.

Best wishes, John

Chapter Nineteen

Errors in the Global Data

Mr. McIntyre also discovered discrepancies in global data, and in April 2005, Ross McKitrick, a McIntyre associate, published information that described the unprofessional activity on the part of the keepers of the keys for global temperature data. (For details, see "What is the Hockey Stick Debate About?" at: http://www.uoguelph.ca/~rmckitri/research/APEC-hockey.pdf.

The big chance for the global warming enthusiasts was created by Michael Mann at Pennsylvania State University. He devised a computer process with which to revisit past world temperatures and, supposedly, to make them more accurate. The result of Mann's work changed previous temperature historical data; it reduced the Earth's temperature magnitude over the past thousand years and increased those in this century, beginning in 1850. He then declared the results of his work proved that the world is in an unprecedented, dangerous situation from increased global warming.

The United Nations Intergovernmental Panel of Climate Change (IPCC) incorporated Mann's questionable study results in its third assessment report in 2001, and that marked the beginning of the world's fears from global warming.

The hockey stick graph depicts the Earth's climate as very stable, with temperatures gradually decreasing from AD1000 to 1900, but it shows temperatures begin to rise "appreciably" from that time on. But you should know that the "appreciable" rise that the "scientists" worry about is only possible to demonstrate if the data analyzed are in anomalies. With CRU data, the percentage increase from the temperature in 1900 (-0.253, 13.747°C, 56.745°F) to that of 2008 (0.327, 14.327°C, 57.789°F), the anomalies show a tremendous 229.2 percent increase. But that isn't a temperature measurement;

when converted to absolute temperature scale, it is only 4.219 percent—a snifter. And when converted to Fahrenheit, it is only 1.84 percent—almost nothing. Using our commercial thermometers and our normal body feelings, we would have noted no change whatsoever.

Mann's revisions contradicted other credible evidence in scientific literature available at the time, but the IPCC's 2001 Third Assessment Report was built comprehensively around Mann's concept, and the resulting furor has been difficult to quell.

Recapturing Climate History of the Past

Before Mann's work, McKitrick shows, the IPCC's Second Assessment in 1995 carried a thousand-year plotted shape of global temperatures that is shown in figure 6.

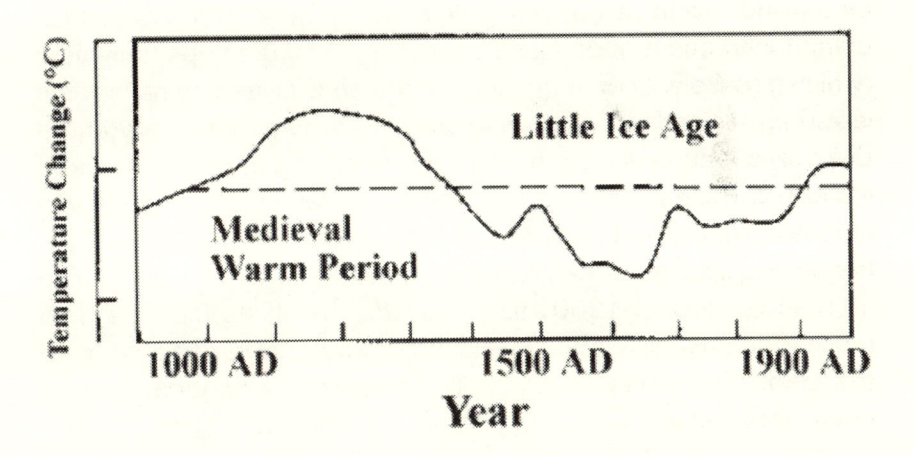

Fig. 6: World Climate History According to IPCC in 1990.

If the beliefs of warming enthusiasts were to endure, the above temperature plot had to be revised in appearance. The plot shows that medieval times were warmer during a time when the levels of CO_2 were obviously much lower. If this were true, then the claims that CO_2 caused global warming could not survive, and the warming enthusiasts' reigning theory would be punctured. (This was a topic discussed in e-mails unearthed by CRU hackers). Unless that early hump was removed, everybody interested would know that global warming, man-made, was obviously a farce. But worst of all, the greatest attraction for climate-science funding, pending world disaster, would disappear.

Greenland Got Its Name during the Medieval Warm Period

Greenland is an arctic area. Today, 85 percent of the country is covered by inland ice, which rises to 9,850 feet above sea level. The most active glacier in the northern hemisphere is found there, in Ilulissat (Jakobshavn), which produces 20 to 25 billion tons of ice annually. Some of the largest icebergs are found around Ilulissat, and these may be up to 330 feet above sea level, even though only 10 to 15 percent of the iceberg is visible above water.

In around 999 AD Eric the Red, with fourteen ships, sailed into Greenland (confirmed by borehole data as about 1000 AD), which offered land and a good climate for crops. The green pasturelands (which it had plenty of at the time) intrigued Eric, and he named the island Greenland. The following year, he brought his family and a few dozen settlers to live there. The climate began to cool at about 1200 AD. (Scientists drilled into the Greenland ice caps to obtain core samples, and the data suggested that the Medieval Warm Period had caused a relatively milder climate in Greenland, lasting from about 800 to 1200 AD. By 1420, the "Little Ice Age" had reached intense levels in Greenland, and the island had become the cold, tundra-packed place we're familiar with today. (www.greenland.com

Chapter Twenty

Recovering Climate Data of the Past

Scientists have legitimate methods to determine the history of climate from the distant past; they use various techniques, and two of them utilize tree rings and ground boreholes. With trees, the distance between the concentric rings of very old trees provides useful information, and with boreholes, the vertical temperature profile of a hole drilled into the ground provides comparable information. A vertical temperature profile of the inside of the hole can be analyzed to obtain an estimate of the historical temperature and climate sequence at the surface.

Mr. Ross McKitrick, in "What is the Hockey Stick Debate About?" mentioned previously, says, "In the mid-1990s the use of ground boreholes as a clue to paleoclimate history was becoming well-established. In 1995 David Deming, a geoscientist at the University of Oklahoma published a study in *Science* magazine that demonstrated the technique by generating a 150-year climate history for North America. Here, in his own words, is what happened next. 'With the publication of the article in *Science*, I gained significant credibility in the community of scientists working on climate change. They thought I was one of them—someone who would pervert science in the service of social and political causes. So one of them let his guard down. A major person working in the area of climate change and global warming sent me an astonishing email that said, 'We have to get rid of the Medieval Warm Period.'"

That may seem sinister to some, but the "scientists" will tell you, if asked, "The 'major person' was not suggesting changes by dishonest means." I say, silly. That "major person" knew, at that early date, that man-made global warming was an economic boon for his profession, and any data—such as warmer temperatures during the medieval period—would be a danger and must be destroyed or covered up.

S. Huang and H. N. Pollack searched the large ground heat flow database of the International Heat Flow Commission of the International Association of Seismology and Physics of the Earth's Interior for measurements suitable for reconstructing an average ground surface temperature history of the earth over the last 20,000 years.[8] Based on a total of 6,144 qualifying sets of measurements

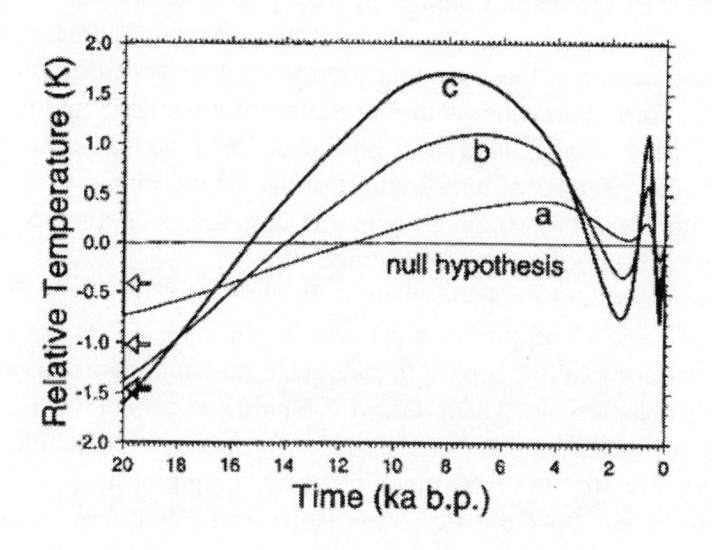

Time is in 1000 years; 0 is today (1997). The three
Curves result from different assumptions. Our interest
Is in the period 2000 years ago and today. The period from
About 1000 to 500 years ago were obviously hotter than today.

Fig. 7

---

[8] "Six Thousand Heat Flow Measurements Can't Be Wrong," updated September 13, 2000, accessed August 1, 2011, http://www.co2science.org/articles/V3/N22/C3.php.

obtained from every continent, they produced a *global* climate reconstruction that they said was "independent of other proxy interpretations [and] of any preconceptions or biases as to the nature of the actual climate history." This effort revealed the existence of a *global* Medieval Warm Period that was as much as 0.5°C (0.9°F) warmer than it was in the late 20th century, as well as a *global* Little Ice Age that was as much as 0.7°C (1.26°F) cooler than it was in the early 1990s. Consequently, and contrary to the climate-alarmist claim that the Medieval Warm Period and Little Ice Age were *neither real nor global*, this study tells a very different story which, because of the massive data base upon which it rests, must be the correct one.

"What is the Hockey Stick Debate About?" *April 4, 2005*

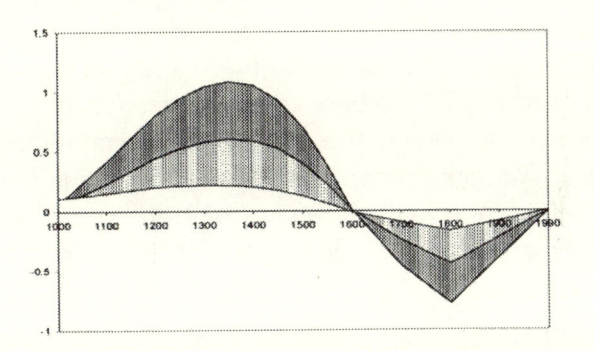

Fig 8

Climate history after 1000 AD, according to ground borehole evidence.
Vertical axis: average anomalies in °C, with range indicating Bayesian probability boundaries.
Source: Huang et al. (1998); data supplied by Huang.

Shaopeng Huang and coauthors published the findings in *Geophysical Research Letters* in 1997. *Nature* magazine essentially ignored the borehole data but the next year published the first Mann hockey stick paper.

Tom Shipley

Henry N. Pollack and Shaopeng Huang, published "Climate Reconstruction From Subsurface Temperatures," in 2000. It reported on the results of studies at that time plus a global study involving six hundred sites and six continents, and said on page 353, under Global Analysis, they ". . . revealed a long mid-Holocene warm interval some 0.2-0.6 K (0.2-.0.6°C) (0.36-1.08°F) above present day temperatures, and another similar but shorter warm interval 500-1,000 years ago. Temperatures then cooled to a minimum of approximately 0.5 K (0.5°C, 0.9°F) below present, about 200 years ago . . ." (http://www.eos.ubc.ca/~mjelline/453website/eosc453/E_prints/AnnRev.28.1.339.pdf)

Yet in July 4, 2008, in *Geophysical Research Letters*, Vol. 35, L13703, Huang, Pollack, and P.-Y. Shen presented results of study data that disagreed with their 1997 and 2000 conclusions: ". . . The reconstructions show the temperatures of the mid-Holocene warm episode some 1-2 K above the reference level, *the maximum of the MWP at or slightly below the reference level*, the minimum of the LIA about 1 K below the reference level, *and end-of-20th century temperatures about 0.5 K above the reference level . . .*" (Emphasis added) For our information, the authors are saying temperatures in the Medieval Warming Period were lower than those at the end of the 20th century. And temperatures at the end of the 20th century were higher than at any time during the MWP, contrary to previous analyses.

Both Huang and Pollock wrote separate articles on new insights into climate change in 2004. It is interesting that the articles were written, and that the tenor of their work changed greatly. Before I would put a bookmarker to hold their place in history, I would wait a little while longer. In the same 2008 dissertation, the authors say, ". . . These reconstructions resolve the warming from the last glacial maximum, the occurrence of mid-Holocene warm period, a MWP and LIA, and the rapid warming of the 20th century, all occurring at times consistent with a broad array of paleoclimatic proxy data. The reconstructions show the temperatures of the mid-Holocene warm period some 1-2 K above the reference level, the maximum of the MWP at or slightly below the reference level, the minimum of the LIA about 1 K below the reference level, and end-of-20th century temperatures about 0.5 K above the reference level. *All of these amplitude estimates are, as with the timing of these episodes, generally consistent with amplitudes estimated from other climate proxies as summarized by Intergovernmental Panel on Climate Change [2007]."* (Emphasis added)

Remember David Deming, the borehole geoscientist at the University of Oklahoma, and the "leading scientist's" plea after Deming was apparently a welcome addition to the "scientist" community? The warming enthusiast said, "We have to get rid of the Medieval Warm Period." As we divulge the knowledge of others as we move along, we may have to ask Huang and Pollock what changed their minds? It is obvious from Mr. Deming's work, this example showing accuracy, and the work of others in this field that borehole data is a dependable tool that can be used to determine the state of world climate during the distant past with satisfactory limits of accuracy. If sufficient borehole data is available from all continents in the world, utilizing suitable depths, the global climate during the past 1,000 to 2,000 years could be determined, without question. Those wanting to "get rid of" the Medieval Warm Period run into the problem that it shows up strongly in borehole studies data. Shortly after Deming's article appeared, a group led by Shaopeng Huang of the University of Michigan completed a major analysis of over 6,000 borehole records from every continent around the world. Their study went back 20,000 years. The portion covering the last millennium is shown in figure 7. The similarity to the IPCC's 1990 graph is obvious. The world experienced a "warm" interval in the medieval era that exceeded twentieth-century changes. The present-day climate appears to be simply a recovery from the cold years of the "Little Ice Age."

Chapter Twenty-One

Several Studies Say Mann and the IPCC are Wrong

The Center for the Study of Carbon Dioxide and Global Change, www.co2science.org, was created to disseminate factual reports and sound commentary on new developments in the worldwide scientific quest to determine the climatic and biological consequences of the ongoing rise in the air's CO_2 content.[9] It meets this objective through weekly online publication of its *CO2 Science* magazine, which contains editorials on topics of current concern and mini-reviews of recently published peer-reviewed scientific journal articles, books, and other educational materials. In this endeavor, the center attempts to separate reality from rhetoric in the emotionally charged debate that swirls around the subject of carbon dioxide and global change. In addition, to help students and teachers gain greater insight into the biological aspects of this phenomenon, the center maintains online instructions on how to conduct CO_2 enrichment and depletion experiments in its "Global Change Laboratory" (located in its "Education Center" section), which allow interested parties to conduct similar studies in their own homes and classrooms.

CO2 Science summarized the results of thirteen studies involving boreholes—from 1997 to 2005. The studies involved both worldwide and regional locations, as follows: Huang and Pollack, 1997 (as discussed earlier); Pollack et al., 1998, eastern North America, central Europe, southern Africa, Australia; Dahl-Jaensen, 1998, Greenland Ice Sheet; Majorowicz et al., 1999, ten Saskatchewan sites; Correia and Safanda, 1999, Lisbon, Portugal; Bodri and Cermak, 1999, Czech Republic; Huang et al., 2000, all continents except Antarctica;

---

[9] "CO2 Science," last updated July 27, 2011, accessed August 1, 2011, http://www.co2science.org/.

Harris and Chapman, 2001, Northern Hemisphere; Demezhko and Shchapov, 2001, Middle Ural Mountains; Romanofsky et al., 2002, Barrow, Alaska; Gonzalez-Rousco et al., investigated techniques for analysis of deep hole measurements; Beltrami et al., northern Quebec; Bodri and Cermac, 2005, Czech Republic, improved methods for borehole analysis.

All of them showed Mann's and the IPCC's conclusions to be incorrect: There was a global medieval warm period, from 1000 to 1350 CE (AD for old timers), and its temperatures equaled, but most results exceeded, the temperatures of the twentieth century. The authors concluded with the observation that "In light of these many eye-opening results of studies of borehole temperature data—which Broecker (2001) describes as one of 'only two proxies' (the other being mountain snowlines) that for time scales greater than a century or two 'can yield temperatures that are accurate to 0.5°C'—it is obvious that climate-alarmist claims of unprecedented CO_2-induced global warming over the past few decades are wholly without significant real-world empirical support. In fact, *they are refuted by it.*"

Something Fishy Here

In early 2000, the IPCC released the first draft of the Third Assessment Review. The 1997 and 1990 borehole studies were ignored, essentially, and the hockey stick scam was the only paleoclimate reconstruction shown in the summary and the only one in the whole report to be singled out for repetition. The borehole data received a brief mention in chapter 2, but the Huang et al. graph was not shown. A small graph of borehole data, taken from another study and based on a smaller sample, was shown in a post-1500 segment, which, conveniently, trended upward.

Huang's study results showed temperatures that followed the path that IPCC had touted prior to the hockey stick revisions, instituted in the 2001 Third Assessment Report. When the IPCC report appeared, the hockey stick version of climate history became solid science. Suddenly it was the "consensus" view, and for the next few years, anyone who publicly questioned the "consensus" was pilloried.

In May 2008, Tad Cronn reported additional details on information previously mentioned here: "Al Gore and global warm-mongers have won many converts with their claim that 2,500 scientific reviewers of the U.N. Intergovernmental Panel on Climate Change's report constitutes a 'consensus' among scientists that man-made warming is destroying Earth." (http://tadcronn.wordpress.com/2008/05/27/global-warming-consensus-31000-scientists-disagree)."

Before proceeding with Mr. Cronn's report, I must introduce information from an article in *Progress in Physical Geography*, April 12, 2010, Entitled *Climate Change: what do we know about the IPCC? pages 10, 11*. It concerns the IPCC's and Al Gore's count of 2500 scientists. Mike Hulme is a professor of climate change in the School of Environmental Sciences at the University of East Anglia. He is the founding director of the Tyndall Centre for Climate Change Research and one of the UK's most prominent climate scientists. Among his many roles in the climate change establishment, Hulme was the IPCC's coordinating lead author for its chapter on "Climate Scenario Development" for its Third Assessment Report and a contributing author of several other chapters. Professor Hulme said this about the claim of 2,500 reviewers and consensus: "Without a careful explanation about what it means, this drive for consensus can leave the IPCC vulnerable to outside criticism. Claims such as '2,500 of the world's leading scientists have reached a consensus that human activities are having a significant influence on the climate' are disingenuous. That particular consensus judgment, as are many others in the IPCC reports, is reached by only a few dozen experts in the specific field of detection and attribution studies; other IPCC authors are experts in other fields."

Mr. Cronn's report continues: "Not only have many of those reviewers made it known that they disagree with the U.N. conclusions, but now there is a petition circulated by Dr. Arthur Robinson, director of the Oregon Institute for Science and Medicine, signed by more than 31,000 scientists who dispute the theory of man-made global warming. The petition states, in part:

"'. . . There is no convincing scientific evidence that human release of carbon dioxide, methane, or other greenhouse gasses is

Man-Made Global Warming?

causing or will, in the foreseeable future, cause catastrophic heating of the Earth's atmosphere and disruption of the Earth's climate. Moreover, there is substantial scientific evidence that increases in atmospheric carbon dioxide produce many beneficial effects upon the natural plant and animal environments of the Earth . . .'

"The 31,000 signers all hold scientific credentials; approximately 9,000 of them hold scientific Ph.D.s. Robinson held a press conference earlier this month. Although members of the media and Congress were invited, attendance was light. Robinson points out that over the past 150 years, scientists have found that global temperatures have been predicted with 79 percent accuracy by the sunspot index, which precedes climate changes by about 10 years. CO_2, by comparison, has been only 22 percent accurate, and that number has rapidly declined in the past decade as temperatures have dipped and CO_2 has continued to rise."

The World Climate Report presented information on March 3, 2005 entitled *Hockey Stick, 1998-2005, R.I.P*, (filed under: Climate History, Paleo/Proxy, Temperature History). "The 'Hockey Stick' representation of the temperature behavior of the past 1000 years is broken, dead. Although already reeling from earlier analyses aimed at its midsection, the knockout punch was just delivered by Nature magazine. Thus the end of this palooka: that the climate of the past millennium was marked by about 900 years of nothing (1000 to 1900 AD) and then 100 years of dramatic temperature rise caused by people. This once-feared icon of global warming purported to show annual average temperature of the Northern Hemisphere for the past 1,000 years. It was derived from the climatic information that is stored in a variety of climate-sensitive or climate "proxy" data records—things such as tree rings, coral banding records, and sediment cores. It's called the 'hockey stick' because its long handle corresponds to 900 years (from 1000 to 1900) of little temperature variation, and its blade represents 100 years (1900 to 1999) of rapid temperature rise . . . The saga of the 'hockey stick' will be remembered as a remarkable lesson in how fanaticism can temporarily blind a large part of the scientific community and allow unproven results to become 'mainstream' thought overnight . . .

". . . The first sign that something amiss with the 'hockey stick' was published in 2003 by Harvard scientists Willie Soon and Sallie

Baliunas. Soon and Baliunas performed a survey of the existing scientific literature concerning the climate of the past 1,000 years and compiled evidence for and against the existence of the Medieval Warm Period (MWP) and Little Ice Age (LIA). They found that overwhelmingly, within the scores of scientific articles that they reviewed, there was strong evidence to support the existence of these well-known climatic episodes that were largely absent from the 'hockey stick' reconstruction. Apparently, the handle of the 'hockey stick'—that part of it which represents natural variation—is too flat.

"Then came the painstaking effort by Steven McIntyre and Ross McKitrick to simply attempt to reproduce the 'hockey stick' using the data and procedures described by Mann and colleagues in their 1998 *Nature* publication. In their professions McIntyre (a mineral consultant) and McKitrick (an economist) had encountered numerous hockey-stick-shaped graphs that were typically used to try to sell an idea based upon some measure of performance. Their experience was that these types of graphs inevitably broke down under careful scrutiny. Familiar with accounting procedures, they decided, out of personal interest, to 'audit' the 'hockey stick' and see if they could recreate it starting from scratch.

"The resulting trials and tribulations of McIntyre and McKitrick make for a truly eye-opening look at the supposed 'openness' of the scientific process. For years they toiled tirelessly in their task, working through countless roadblocks erected by the 'hockey stick's' original creators, and documenting an embarrassing number of errors in the original procedure including inaccurate data descriptions, insufficient methodological details, data compilation errors, data handling mistakes, and questionable statistical techniques. While no individual mistake was likely sufficient enough in and of itself to throw into question the 'hockey stick,' taken together, the list of errors indicate a certain lack of rigor and attention to detail by the 'hockey stick's' creators. Their efforts are detailed in two scientific articles (McIntyre and McKitrick, 2003; 2005), in an upcoming book chapter, and in McIntyre's personal web page. Additionally, the *Wall Street Journal* chronicled much of this activity in a front page article on February 14, 2005.

"The third dissenting voice was that of Jan Esper and colleagues in 2004. Esper is an expert in climate reconstructions based upon tree-ring records (the primary type of proxy data relied upon by Mann et al. in creating the 'hockey stick'). It turns out that one must be careful when using tree rings to reconstruct long-term climate variability because as the tree itself ages, the widths of the annual rings that it produces changes—even absent any climatic variations. This growth trend needs to be taken into account when trying to interpret any climate data contained in the tree-ring records . . . Esper et al. point out that this could be one likely reason why the handle of the 'hockey stick' is so flat—it lacks the centennial-scale variations that were lost in the standardization of its primary data source. Using an alternative technique that attempted to preserve as much of the information about long-term climate variations as possible from historical tree-ring records, Esper and colleagues derived their own annual Northern Hemisphere temperature reconstruction. The result was a 1,000-yr temperature history in which the LIA and the MWP are much more pronounced than the 'hockey stick' reconstruction—more evidence that the 'hockey stick' underestimates the true level of natural climate variation . . ."

"The chorus of dissent grew louder with the publication of a paper by Hans von Storch and colleagues in *Science* in late 2004. Von Storch was interested in how well the temperature reconstruction methodology used in producing the 'hockey stick' actually worked. In order to investigate this, he used a climate model, run with historic changes in solar output and volcanic eruptions to produce a temperature record for the past 1,000 years." [Using a methodology similar to the Mann etal.'s] ". . . Von Storch's research team found that the techniques used to construct the 'hockey stick' vastly underestimated the true level of variability in the known (modeled) temperature record.

". . . And [finally], with the publication of a paper in *Nature* magazine in early 2005 by Anders Moberg and colleagues, it's all over for the hockey stick. Recognizing that different kinds of proxy temperature records may be more appropriately related to climatic variations at different time scales, Moberg applied a statistical technique called 'wavelet analysis' that allows each proxy [tree rings, lake and ocean sediments, etc.] to explain temperature variations on

Tom Shipley

a timescale that it was most sensitive to." Moberg's reconstruction showed ". . . strong MWP and LIA signals. The natural variation of temperatures in the Moberg reconstruction is two to three times that of the Mann et al. 'hockey stick.' Again, the handle of the 'hockery stick' was found to be too flat . . ."

The *World Climate Report* concludes, ". . . But, the 'hockey stick' was remarkable. And as such, it will be remembered as a remarkable lesson in how fanaticism can temporarily blind a large part of the scientific community and allow unproven results to become mainstream thought overnight. The embarrassment that it caused to many scientists working in the field of climatology will not be soon forgotten. Hopefully, new findings to come, as remarkable and enticing as they may first appear, will be greeted with a bit more caution and thorough investigation before they are widely accepted as representing the scientific consensus . . ."

Chapter Twenty-Two

Repetitive Cycle Patterns

Girma Orssengo, MASc, PhD, and global thinker, dissents from the hockey stick results, takes a different tack, and analyzes and compares temperature data from 1810 to 1910 (before the dawn of industrialization) with that from 1850 to 2008 (CRU data for that entire period).[10] The result is explained this way:

> This linear warming of 0.47 deg C/100 years, two centuries ago, is of similar magnitude to that of the last century's value of 0.44 deg C/100 years. There was no significant change in the linear anomaly in the previous two centuries . . . the linear warming of the last century was not caused by human emission of CO2.
> Science is about the data. Science is not about consensus or authority.
> The linear global warming of the last century was [also] similar to that of two centuries ago—the oscillating warming by 0.67 deg C from 1976 to 1998 is as natural as the oscillating cooling by similar amount from 1878 to 1911.
> There is no shift in mean global temperature anomaly in the last century as a result of CO_2 emission. None.

There is no question, as shown previously, that global temperatures have risen during the past 158 years—but Orssengo is showing us that humans' puny pilgrimages did not contribute to the puny rise. The rise during the last one hundred years, involving heavy production and consumer use, was essentially the same as

---

[10] Girma Orssengo, "CO_2 Driven Global Warming Is Not Supported by the Data," accessed August 1, 2011, http://orssengo.com.

the preceding one hundred years, in which the human contribution was nonexistent.

This entire exercise, performed by M. E. Mann, R. S. Bradley, and M. K. Hughes, was designed to fit the global temperature pattern needed by struggling "scientists" and their employers in their search for funding. It has been done before, but, we have to hand it to them, never as skillfully. And, never before, it appears, had it been so helped along by media.

Do I Dare Say Once Again

I say, regardless of the number of doctoral degrees a scientist holds or the length of his or her experience in reading and analyzing data from thermometers and sensors, nobody knows, definitively, where the temperature is going in the future. As any thinking person knows, scientists can't tell us with any degree of accuracy what the temperature next week will be. We can guess; we can theorize; we can hypothesize. But we can't know.

Why are we getting a barrage of incorrect information that is touting global warming? It's simple: The federal government and others have dangled a pot full of money that they don't have that is dedicated to resolving a problem that we don't have. Former vice president Al Gore, a politician with no experience in science, engineering, or running a business (he didn't even finish college) is leading the charge, and those in line behind him, all seeking their share of the loot, are beating the drum to assure that the money keeps flowing.

As discussed previously, the media can be very little help, but data are available. I was alerted by the *Wall Street Journal,* and two commentators, Frank Beckman and Rush Limbaugh, on our local radio station, WJR. But I have seen little that presents the tiny range of temperatures involved, mostly too small to read on our thermometers, in terms that the American people can understand.

I think, as indicated previously, the popular media, used primarily by people who aren't paying enough attention, lack knowledge. (This lack of attention is no reflection on the people; they are just busy running their own lives.) Too many of the popular media's

Man-Made Global Warming?

reporters have only journalism degrees, and they only know how to report, skillfully, on what somebody says or writes. I think they don't realize that factual data, easily obtained, can be used to bulk up their reports and make them more complete and true. Or maybe, and I suspect this plays the largest part, they think reports on the argument between factions is more important—it makes for greater reader interest.

Chapter Twenty-Three

At this point in the story, what do you think? Is it just a series of mistakes or fraudulent intent? I will give you my reaction, reached two years ago: If these alarmed "scientists" had worked for any company that I have been associated with over the years, they would never have reached this stage in their careers. I thought the great problems with university professionals existed only in the philosophical and non-technological sectors of higher education, but I now know they have spilled over into the technical sector also.

Think about the lack of accuracy that is now obvious (in one or more degrees Fahrenheit, not increments as small as 0.318 or less about which we are arguing) throughout the entire climate–and temperature-monitoring system of the United States–reputed to be the best. Then think about your tax dollars that your Congress members are authorizing, which your president is in the process of spending. And look at the miserable little representatives, reputed to be automobiles, that cruise the streets–head down, tail up–and all makes look alike. To my eighty-seven-year-old eyes, they all look alike and are the ugliest I have ever seen.

More Earned Misery for the Errant "Scientists"

And then hackers struck at the CRU (U. of E. Anglia, UK). The following information was reported by Anthony Watts in November 2009, (http://wattsupwiththat.com/2009/11/19/breaking-news-story-hadley-cru-has-apparently-been-hacked-hundreds-of-files-released).

Man-Made Global Warming?

"I'm currently traveling and writing this from an airport, but here is what I know so far: An unknown person put postings on some climate skeptic websites that advertised an FTP file (a common method of transferring files via the Internet from one computer to another): on a Russian FTP server here is the message that was placed on the Air Vent today:

> 'We feel that climate science is, in the current situation, too important to be kept under wraps.
>
> 'We hereby release a random selection of correspondence, code, and documents.'

"The file was large, about 61 megabytes, containing hundreds of files. It contained data, code, and emails from Phil Jones at CRU to and from many people. I've seen the file, it appears to be genuine and from CRU. Others who have seen it concur—it appears genuine. There are so many files it appears unlikely that it is a hoax. The effort would be too great."

The *New York Times*, in writing about the hacking, commented in its usual liberal fashion: "The evidence pointing to a growing human contribution to global warming is so widely accepted that the hacked material is unlikely to erode the overall argument. However, the documents will undoubtedly raise questions about the quality of research on some specific questions and the actions of some scientists."

The Guardian, November 20 edition; The Wall Street Journal, Novermber 23; The Telegraph, November 20; and many others reported on the hacked e-mails.

On November 20, 2009, Ed Morrissey, (http://hotair.com/archives/2009/11/20/do-hacked-e-mails-show-global-warming-fraud/), wrote,

> ... A 62 megabyte zip file (a computer file whose contents are compressed for storage or transmission), containing around 160 megabytes of emails, pdfs [a special computer

file format) and other documents, has been confirmed as genuine by the head of the University of East Anglia's Climate Research Unit, Dr. Phil Jones.

In an exclusive interview with *Investigate* magazine's TGIF Edition, Jones confirms his organization has been hacked, and the data flying all over the internet appears to have come from his organization. 'It was a hacker. We were aware of this about three or four days ago that someone had hacked into our system and taken and copied loads of data files and emails.'

One of the most damning e-mails published comes from Dr. Jones himself. In an e-mail from almost exactly ten years ago, Jones appears to discuss a method of overlaying data of temperature declines with repetitive, false data of higher temperatures:

> From: Phil Jones
> To: ray bradley, mann@[snipped], mhughes@[snipped]
> Subject: Diagram for WMO Statement
> Date: Tue, 16 Nov 1999 13:31:15 +0000
> Cc: k.briffa@[snipped], t.osborn@[snipped]
>
> Dear Ray, Mike and Malcolm,
>
> Once Tim's got a diagram here we'll send that either later today or first thing tomorrow. I've just completed Mike's Nature trick of adding in the real temps to each series for the last 20 years (ie from 1981 onwards) and from1961 for Keith's to hide the decline. Mike's series got the annual land and marine values while the other two got April-Sept for NH land N of 20N. The latter two are real for 1999, while the estimate for 1999 for NH combined is +0.44C wrt 61-90. The Global estimate for 1999 with data through Oct is +0.35C cf. 0.57 for 1998.
>
> Thanks for the comments, Ray.

Cheers, Phil
Prof. Phil Jones
Climatic Research Unit

Jones told *Investigate* that he couldn't remember the context of "hide the decline," and that the process was a way to fill data gaps rather than mislead. But when scientists talk about "tricks" in the context of hiding data, it certainly seems suspicious.

[Note: Anybody who has spent time working with CRU's temperature data knows precisely what Phil Jones was talking about: He was doctoring temperature data to fit his pattern and that of other global warming enthusiasts to show increased temperatures at the proper, necessary places. (Figure 1 shows a decrease in temperatures lasting from about 1950 to 1980; he needs to increase those and those from 1981 to make global warming more believable.) Andrew Bolt points to a couple of other suspicious entries in the database, as well, for the *Herald-Sun*. For instance, next we have scientists discussing how to delete inconvenient data in order to emphasize other data that supports their conclusions:]

Morrisey continues:

> From: Tom Wigley [. . .]
> To: Phil Jones [. . .]
> Subject: 1940s
> Date: Sun, 27 Sep 2009 23:25:38–0600
> Cc: Ben Santer [. . .]
> Phil,
>
> Here are some speculations on correcting SSTs [sea surface temperatures] to partly explain the 1940s warming blip. If you look at the attached plot you will see that the land also shows the 1940s blip (as I'm sure you know).

Tom Shipley

So, if we could reduce the ocean blip by, say, 0.15 degC, then this would be significant for the global mean—but we'd still have to explain the land blip. I've chosen 0.15 here deliberately. This still leaves an ocean blip, and i think one needs to have some form of ocean blip to explain the land blip (via either some common forcing, or ocean forcing land, or vice versa, or all of these). When you look at other blips, the land blips are 1.5 to 2 times (roughly) the ocean blips—higher sensitivity plus thermal inertia effects. My 0.15 adjustment leaves things consistent with this, so you can see where I am coming from.

Removing ENSO [El Niño/La Niña-Southern Oscillation, or ENSO] does not affect this.

It would be good to remove at least part of the 1940s blip, but we are still left with "why the blip".

Let me go further. If you look at NH [Northern Hemisphere] vs. SH [Southern Hemisphere] and the aerosol effect (qualitatively or with MAGICC) then with a reduced ocean blip we get continuous warming in the SH, and a cooling in the NH—just as one would expect with mainly NH aerosols.

The other interesting thing is (as Foukal et al. note—from MAGICC) that the 1910-40 warming cannot be solar. The Sun can get at most 10% of this with Wang et al solar, less with Foukal solar. So this may well be NADW, as Sarah and I noted in 1987 (and also Schlesinger later). A reduced SST blip in the 1940s makes the 1910-40 warming larger than the SH (which it currently is not)—but not really enough.

So . . . why was the SH so cold around 1910? Another SST problem? (SH/NH data also attached.)

This stuff is in a report I am writing for EPRI, so I'd appreciate any comments you (and Ben) might have.

Tom.

Hmmm. Sounds like "hid[ing] the data" once again. And here we have them privately admitting that they can't find the global warming that they've been predicting.

[Note: Refer to figure 1, CRU data, and you will find a peak in temperature in 1940, and then the global temperatures decreased until about 1980, when they began to rise sharply again. The e-mail exchange appears to say that decreasing the temperatures during the '40s would eliminate the blip (the increase in temperature), which decreased until 1980, while emitted CO_2 was exploding. The decreasing temperature period while CO_2 was increasing substantially upsets the CO_2-caused warming theory.]

> From: Kevin Trenberth
> To: Michael Mann
> Subject: Re: BBC U-turn on climate
> Date: Mon, 12 Oct 2009 08:57:37−0600
> Cc: Stephen H Schneider, Myles Allen, peter stott, "Philip D. Jones", Benjamin Santer, Tom Wigley, Thomas R Karl, Gavin Schmidt, James Hansen, Michael Oppenheimer

> Hi all

> Well I have my own article on where the heck is global warming? We are asking that here in Boulder where we have broken records the past two days for the coldest days on record. We had 4 inches of snow. The high the last 2 days was below 30F and the normal is 69F,

Tom Shipley

and it smashed the previous records for these days by 10F. The low was about 18F and also a record low, well below the previous record low.

This is January weather (see the Rockies baseball playoff game was canceled on Saturday and then played last night in below freezing weather).

The fact is that we can't account for the lack of warming at the moment and it is a travesty that we can't. The CERES data published in the August BAMS 09 supplement on 2008 shows there should be even more warming: but the data are surely wrong. Our observing system is inadequate.

Kevin

In response to that, Morrissey asks :

Do scientists use data to test theories, or do they use theories to test data? Scientists will claim the former, but here we have scientists who cling to the theory so tightly that they reject the data. That's not science; it's religious belief.

Dr. Jones has confirmed that these e-mails are genuine. Whether the work represented by these scientists is as genuine seems to be under serious question. Tim Blair says, 'The fun is officially underway.'

Update: These e-mails may explain this:

Global warming appears to have stalled. Climatologists are puzzled as to why average global temperatures have stopped rising over the last 10 years. Some attribute the trend to a lack of sunspots, while others explain it through ocean currents.

At least the weather in Copenhagen is likely to be cooperating. The Danish Meteorological Institute predicts that temperatures in December, when the city will host the United Nations Climate Change Conference, will be one degree above the long-term average.

Otherwise, however, not much is happening with global warming at the moment. The Earth's average temperatures have stopped climbing since the beginning of the millennium, and it even looks as though global warming could come to a standstill this year.

Or maybe it didn't exist at all, except when scientists at Hadley were "hid[ing] the decline[s].

Update II: This follows on a more mundane controversy over competence at Hadley that erupted in September:

A scientific scandal is casting a shadow over a number of recent peer-reviewed climate papers. At least eight papers purporting to reconstruct the historical temperature record times may need to be revisited, with significant implications for contemporary climate studies, the basis of the IPCC's assessments. A number of these involve senior climatologists at the British climate research centre CRU at the University East Anglia. In every case, peer review failed to pick up the errors.

Chapter Twenty-Four

Massachusetts Institute of Technology's online newspaper, the *Tech*, published Matthew Davidson's December 1, 2009, article, "'Hackers' Reveal Corrupt Science at Climate Research Unit." Davidson weighs in with comments:

> Over the past few weeks anonymous "hackers" entered the computer systems of the Climate Research Unit of the University of East Anglia in the UK. This intrusion has been confirmed by the university and at least some of the data leaked to Wikileaks.org have been confirmed as authentic by officials at the CRU. Among the data were hundreds of e-mails and source code files which describe a shameful corruption of the scientific process.
>
> Many corporate media outlets have refused to report on this story. Indeed much of the biased reporting recently put forth as journalism by CNN, *The New York Times*, etc. has presented anthropogenic global warming as a foregone conclusion. In reality, there is a great deal of disagreement among scientists on the subject. Looking beyond these media giants to the independent journalism being conducted by bloggers across the web, one can find a great deal of evidence contradicting the foregone conclusions of corporate media and revenue-hungry politicians.
>
> Warwick Hughes, an Australian climatologist with a skeptical view of anthropogenic global warming, contacted Phil Jones, the director of CRU in 2000, to ask about some inconsistencies he saw in Jones' work. He asked for data gathered using government funds, which he is entitled to view under the UK's Freedom of Information Act of 2000. After a series of e-mails, Phil Jones replied: "Why should I make

the data available to you, when your aim is to try and find something wrong with it." Warwick Hughes was repeatedly denied access to this data in a move to conceal materials and methods which should be open to any honest scientific debate and discussion.

When other scientists began asking for data from the CRU using FOIA requests, Phil Jones began asking scientists to delete e-mail records: "Mike, Can you delete any emails you may have had with Keith re AR4? Keith will do likewise. Can you also email Gene and get him to do the same? We will be getting Caspar to do likewise." If these e-mails had been requested under FOIA, deletion of these records was a likely a criminal act.

Also uncovered was an e-mail that expressed cheer at the death of a scientist who published reviews skeptical of anthropogenic global warming and communicated legal advice on how intellectual property rights may be used to conceal data: "Subject: John L. Daly Dead; Mike, In an odd way this is cheering news! One other thing about the CC paper—just found another email—is that McKitrick says it is standard practice in Econometrics journals to give all the data and codes!! According to legal advice IPR [intellectual property rights] overrides this."

Another e-mail from Phil Jones discusses preventing the opposing views of Chris De Freitas from being published in the IPCC Fourth Assessment Report: "The other paper by MM is just garbage—as you knew. De Freitas again. Pielke is also losing all credibility as well by replying to the mad Finn as well—frequently as I see it. I can't see either of these papers being in the next IPCC report. K and I will keep them out somehow—even if we have to redefine what the peer-review literature is!"

E-mails of other scientists at the CRU were also released. Tim Osborn of the CRU discusses truncating data to "hide" declining temperatures. An e-mail from Prof. Michael Mann of Penn State University to Tim Osborn stated that results which support critics of global warming shouldn't be shown to others. Dr. Kevin Trenberth of the National Center for

Atmospheric Research admits that he cannot account for the current lack of global warming: "The fact is we can't account for the lack of warming at the moment and it is a travesty that we can't."

These communications reveal a trail of manipulation and concealment of data that would not support the theory of anthropogenic global warming. This is shameful and cannot be ignored by the scientific community. This corruption must be investigated and the individuals responsible must be tried for any illegal acts committed.

If we sit by and allow lawmakers to pass and approve legislation based on falsified data and incorrect theories, we are all to blame for the needless negative effects that the new laws will have on our lives.

In a carbon-constrained world, artificial scarcity of a government apportioned commodity will cause some industrial processes in certain areas to be unprofitable, while processes which pollute even more remain profitable in other areas of the world. Prices will rise in some areas and fall in others. Factories will be relocated in order to seek out the most acceptable business environments. The quality of life for individuals living in developed nations is likely to decline greatly as we are asked to pay more for the energy and food we consume. We may eventually be forced to agree to home energy audits and asked to pay fees to the government for items that consume energy in unapproved ways, such as incandescent light bulbs. Under the Copenhagen treaty, these policies will be forced upon us not by our elected officials, but by unelected bureaucrats at the UN and its offshoots.

As we pay these higher prices and fees, the money will flow into government hands. Many lawmakers see climate legislation as a revenue goldmine, and they are very keen on passing this legislation and providing scientists with plenty of motivation (grants) for justifying the theory of anthropogenic global warming.

We must tell our elected representatives that we will not sit idly by while they pass legislation based on compromised data. We must tell them that MIT stands for honesty in science and that we demand a thorough investigation of this matter.

Now comes the question: Do the hacked e-mails prove that man-made global warming is a hoax? No, the revelation provided by the hacked e-mails is one more example of incompetence—or fraudulent behavior—of the "scientists" involved. If you have read this far, you have already realized, pages ago, that the "man-made" theory, developed by what turned out to be a small group of irresponsible scientists, has no proof behind it. And all scientific theories require proof before they can be accepted as "settled science." The requirement goes beyond a vote and a show of hands.

What Do TV Weathercasters Think?

NewsMax, in its July 2011 issue, reported "*TV Weathermen Pour Cold Water on Global Warming. It* used to be that the hot—button issue of global warming was divided along philosophical lines: liberals on one side, conservatives on the other . . . There are also doubters among the people paid to follow our climate every day—America's weather forecasters . . . In a recent study by George Mason University and the University of Texas at Austin, climatologists were shocked to discover that only about half of TV weather casters believe that global warming is occurring at all, and less than a third believe that human activities are causing global warming. In fact, 29 percent agree that the entire global warming issue is a 'scam.'"

The Oslo committee, with people who have been paying attention, has become a subject for laughter.

Chapter Twenty-Five

Now that we have looked at relatively recent events, let's take one long look back—a long way back—at a much wider view of the climate history of our world. I think you will realize, as I did, that some of our scientists haven't made much of a study of world climate history. You have to be a buff to understand much of this. One up coming example: Plate tectonic. Plate tectonics theorizes that the surface of the Earth is comprised of a number of rigid plates or large crustal slabs that are constantly moving relative to each other, that explain observed continental drift.

CLIMATE HISTORY

Christopher R. Scotese's PALEOMAP Project gives us the history of the *plate tectonic* development of the ocean basins and continents, as well as the changing distribution of land and sea during the past 1,100 million years and the corresponding climate history.[11]

The University of California Museum of Paleontology tells us
The main features of the *plate tectonics* theory are:

- The Earth's surface is covered by a series of crustal plates.
- The ocean floors are continually moving, spreading from the center, sinking at the edges, and being regenerated.
- Convection currents beneath the plates move the crustal plates in different directions.

---

[11] Christopher R. Scotese, "PALEOMAP Project," last updated April 2, 2002, accessed August 1, 2011, www.scotese.com/climate.htm.

- The source of heat driving the convection currents is radioactivity deep in the Earth's mantle.[12]

The PALEOMAP Project provides climate information from its studies, covering the world's historical climate variations going back millions of years. An immense amount of historical information is provided here, but I found the climate animation to be most interesting.

While at the site, from the column at left, select "Animation" and then scroll to the bottom of the next screen to "Paleoclimate Animations" and select "Climatic Change: Paleoclimate VR." At the bottom left of the screen is a partial map of the world. Move it up. Place the cursor at the bottom left of the map, depress the mouse, and drag very slowly from left to right. As you do, the map will revert back in time from 000 years, in increments of 10 million years, to show the world as the continents and seas were formed and located 750 million years ago.

Then, if you liked that, return to the previous location, and just above the previous selection, select "Plate Motions 250 Million Years into the Future: Future VR." In the new screen, note the map, and from the left, depress the mouse and drag slowly to the right and note how the continents and seas change in appearance and location as the Earth advances 250 million years into the future.

Please, please—do not tell Al Gore about this. He is not the studious type (he didn't finish college, you know; he dropped out of Vanderbilt law school to run for Congress), so he will not have read about these changes, theoretical as they may be, destined for the future. If he hears that the continents are moving, he will start a movement of his own—perhaps joined in by the "scientists" of global-warming fame—and there is no telling what our politicians will begin to do to stall the inevitable.

---

[12] "The Mechanism behind Plate Tectonics," accessed August 1, 2011, http://www.ucmp.berkeley.edu/geology/tecmech.html.

Tom Shipley

Back to Serious Things—Past Climate

The PALEOMAP Project information includes this excerpted information:

> We can determine the past climate of the Earth by mapping the distribution of ancient coals, desert deposits, tropical soils, salt deposits, glacial material, as well as the distribution of plants and animals that are sensitive to climate, such as alligators, palm trees & mangrove swamps . . .
>
> The Earth's climate is primarily a result of the redistribution of the Sun's energy across the surface of the globe. It is warm near the Equator and cool near the Poles. Wetness, or rainfall, also varies systematically from the equator to the pole. It is wet near the equator, dry in the subtropics, wet in the temperate belts and dry near the poles. Certain kinds of rocks form under specific climatic conditions. For example coals occur where it is wet, bauxite occurs where it is warm and wet, other types occur where it is warm and dry, while others occur where it is wet and cool. The ancient distribution of those and other rock types can tell us how the global climate has changed through time and how the continents have travelled across climatic belts.
>
> The maps showing the Earth's ancient climate were made by mapping the past positions of the continents and plotting on these maps the distribution of rock types that form in specific climatic belts.

ICE HOUSE or HOT HOUSE?

The site provides comprehensive information on the methods used in their studies as verification of the details provided in figure 9. Note the hot and cold distributions that occurred long before human beings arrived with world-disturbing habits and creative messiness. It is also interesting to note the vast changes in regional climate compared to that of today, and some of the comments of the historians.

Man-Made Global Warming?

For the last 5 million years the Earth has been in a major Ice Age. There have been only a few times in Earth's history when it has been as cold as it has been during the last 5 million years.

Ice Age: When the Earth is in its "Ice House" climate mode, there is ice at the poles. The polar ice sheet expands and contracts because of variations in the Earth's orbit (Milankovitch cycles). The last expansion of the polar ice sheets took place about 18,000 years ago.

The following climate period names are shown on the graphic:

Miocene time climate: The climate was similar to today's climate, but warmer. Well-defined climatic belts stretched from Pole to Equator; however, there were palm trees and alligators in England and Northern Europe. Australia was less arid than it is now.

Oligocene time climate: During the Oligocene period, ice covered the South Pole but not the North Pole. Warm Temperate forests covered northern Eurasia and North America.

Early Eocene: During the Early Eocene alligators swam in swamps near the North Pole, and palm trees grew in southern Alaska. Much of central Eurasia was warm and humid.

Late Cretaceous; a ten-mile wide comet, the Chicxulub, struck between North and South America. The impact caused global climate changes that killed the dinosaurs and many other forms of life. By the Late Cretaceous the oceans had widened, and India approached the southern margin of Asia. During the Cretaceous the South Atlantic Ocean opened, India separated from Madagascar and raced northward on a collision course with Eurasia, North America was connected to Europe, and Australia was still joined to Antarctica.

The Precambrian, the last at the bottom of the graphic, began 650 million years ago. A map illustrates the break-up of the supercontinent, Rodinia, which formed 1,100 million years ago. The Late Precambrian was an "Ice House" World, much like the present day.

Tom Shipley

During the last 2 billion years the Earth's climate has alternated between a frigid "Ice House," like today's world [note: the project refers to today's world as "frigid"], and a steaming "Hot House," like the world of the dinosaurs. This chart [figure 9] shows how global climate has changed through time.

Man-Made Global Warming?

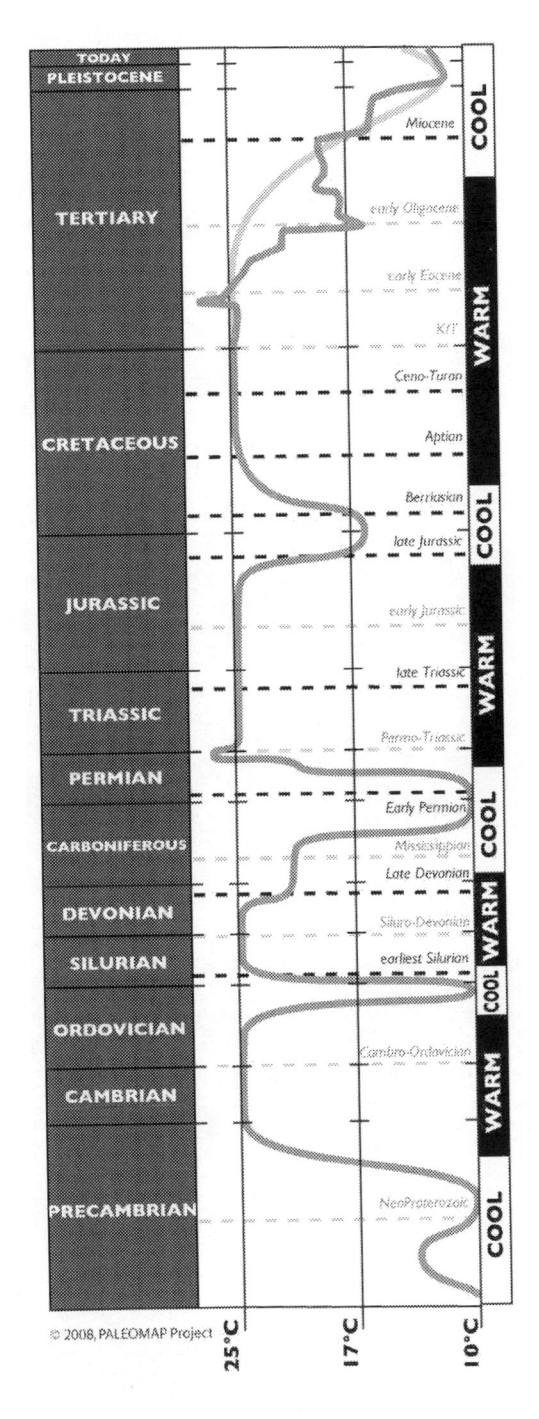

Fig. 9

Tom Shipley

Global Temperatures approaching and leaving the coldest year, 1911.
The data show global temperature anomalies, with the final averages for anomalies and degrees Fahrenheit.

Year	Jan	Feb	Mar	Apr	May	June	July	Aug	Sept	Oct	Nov	Dec	AVG Anomaly	Fahrenheit Degrees F
1908	-0.48	-0.5	-0.65	-0.57	-0.55	-0.51	-0.53	-0.55	-0.49	-0.62	-0.62	-0.59	-0.554	56.203
1908	47	49	51	51	49	48	50	51	47	51	51	53		
1909	-0.58	-0.58	-0.69	-0.65	-0.62	-0.55	-0.6	-0.4	-0.4	-0.5	-0.47	-0.67	-0.559	56.194
1909	51	50	51	51	50	51	54	52	51	52	52	52		
1910	-0.42	-0.58	-0.49	-0.48	-0.56	-0.56	-0.49	-0.51	-0.49	-0.55	-0.69	-0.7	-0.544	56.221
1910	54	52	50	49	52	53	50	52	52	54	53	52		
1911	**-0.59**	**-0.75**	**-0.7**	**-0.73**	**-0.63**	**-0.58**	**-0.54**	**-0.52**	**-0.51**	**-0.52**	**-0.45**	**-0.36**	**-0.573**	**56.169**
1911	56	55	55	52	52	50	53	57	47	52	52	52		
1912	-0.42	-0.4	-0.46	-0.4	-0.44	-0.39	-0.5	-0.59	-0.6	-0.68	-0.56	-0.52	-0.496	56.307
1912	53	53	53	54	55	54	55	55	54	55	54	55		
1913	-0.52	-0.56	-0.58	-0.5	-0.56	-0.56	-0.5	-0.44	-0.47	-0.49	-0.35	-0.3	-0.485	56.327
1913	57	61	61	56	58	59	58	59	60	61	58	58		
1914	-0.16	-0.32	-0.38	-0.43	-0.35	-0.34	-0.38	-0.28	-0.35	-0.24	-0.27	-0.32	-0.318	56.628
1914	57	59	58	57	56	57	55	55	46	47	49	49		
1915	-0.22	-0.13	-0.33	-0.16	-0.32	-0.31	-0.2	-0.19	-0.22	-0.31	-0.23	-0.33	-0.246	56.757
1915	47	47	47	48	47	46	47	47	47	47	45	49		

Figure 10

Man-Made Global Warming?

Global temperature approaching and leaving the hottest year, 1998, in 158 years. Note that the leaving temperatures drop. The temperature drop from 1998 to 2008, in degrees Fahrenheit, does not appear to be very precipitous. However, that drop represents 18.2% of the total rise, 1.98 degrees, seen in 158 years. Is this a signal, as some scientists say, of a coming ice age?

Global Temperatures, Degrees Celsius, with final averages converted to Degrees Fahrenheit (hadcrt3vgl)

Year	Jan	Feb	Mar	Apr	May	June	July	Aug	Sept	Oct	Nov	Dec	AVG Anomaly	Fahrenheit Degrees F
1995	0.347	0.45	0.285	0.23	0.174	0.283	0.279	0.305	0.212	0.241	0.268	0.165	0.27	57.686
1995	82	82	83	82	80	80	81	82	82	81	81	82		
1996	0.067	0.244	0.125	0.095	0.176	0.161	0.178	0.176	0.092	0.087	0.08	0.175	0.138	57.448
1996	83	83	83	80	79	80	80	80	81	80	80	82		
1997	0.157	0.244	0.26	0.198	0.255	0.365	0.373	0.404	0.455	0.482	0.448	0.527	0.347	57.825
1997	82	82	82	81	81	80	80	81	80	79	80	81		
1998	**0.484**	**0.732**	**0.519**	**0.602**	**0.567**	**0.576**	**0.65**	**0.612**	**0.399**	**0.406**	**0.343**	**0.426**	**0.526**	**58.147**
1998	81	81	81	79	80	78	79	79	78	79	79	80		
1999	0.363	0.534	0.286	0.318	0.246	0.266	0.281	0.251	0.274	0.239	0.223	0.341	0.302	57.744
1999	80	80	80	79	78	78	78	80	79	80	80	81		
2000	0.212	0.361	0.332	0.445	0.267	0.249	0.26	0.337	0.31	0.21	0.159	0.183	0.277	57.699
2000	82	82	80	79	78	78	77	79	77	78	78	79		
2001	0.329	0.289	0.474	0.426	0.399	0.416	0.453	0.498	0.403	0.376	0.489	0.323	0.406	57.931
2001	79	80	80	78	77	78	78	80	78	78	79	80		
2002	0.57	0.594	0.586	0.443	0.432	0.458	0.462	0.412	0.412	0.362	0.398	0.327	0.455	58.019
2002	79	81	81	79	79	77	78	79	78	79	81	80		

107

Tom Shipley

Year	Jan	Feb	Mar	Apr	May	June	July	Aug	Sept	Oct	Nov	Dec	AVG Anomaly	Fahrenheit Degrees F
2003	0.515	0.424	0.415	0.404	0.437	0.435	0.454	0.511	0.497	0.549	0.418	0.517	0.465	58.037
2004	81	82	80	79	79	79	78	79	79	78	79	81		
2005	0.452	0.36	0.489	0.532	0.474	0.506	0.531	0.498	0.494	0.501	0.489	0.369	0.475	58.055
2005	80	81	80	78	78	79	80	81	81	80	81	82		
2006	0.298	0.431	0.384	0.364	0.349	0.447	0.446	0.478	0.421	0.473	0.435	0.529	0.421	57.958
2006	81	82	83	81	80	81	81	81	82	82	81	82		
2007	0.622	0.509	0.437	0.467	0.372	0.373	0.392	0.362	0.401	0.361	0.263	0.23	0.399	57.918
2007	81	81	81	81	80	81	81	82	82	81	81	82		
2008	0.05	0.199	0.478	0.285	0.282	0.312	0.405	0.403	0.37	0.447	0.399	0.325	0.327	57.789
2008	82	83	84	83	82	82	82	83	82	83	82	82		

Figure 11

The Author

Tom Shipley is an electrical engineer by study and past profession. He was born in Kingsport, Tennessee, served in the US Navy during World War II, and was a 1950 graduate of Virginia Polytechnic Institute in Blacksburg, Virginia, with a BSEE degree (with honors).

He joined the General Electric Company in September of 1950 and served in engineering, the corporate and field sales function, and as a corporate, division, and district application engineer. From 1960 to 1965, he functioned in General Electric's Specialty Control Department as a product planner and application engineer—first for computer numerical controls for machine tools and then for computer-operated test and inspection systems.

In 1965, he joined Monarch Machine Tool Company in Sidney, Ohio, as manager of numerically controlled machine sales and in 1969 was promoted to international sales manager. Prior to forming the Machine Tool Sales Company in 1976, he served as vice president of marketing for AA Gage Division of US Industries, Inc.

To illustrate his credentials, the following portion of his biography was presented in 1989 when he began writing a regular column for *Metalworking Production and Purchasing* magazine:

> Tom Shipley is a prolific writer and has been widely published by the trade press for general industrial, electrical, electronic, and metalworking sectors. He has spoken on a wide range of subjects for various associations—Society of Mechanical Engineers (SME), Institute of Electrical & Electronic Engineers (IEEE), Numerical Control Society (NCS), Abrasive Engineering Society (AES), and others. He was a featured speaker at the first Systems Engineering Conference held in New York City in 1964.

Tom Shipley

He has been a member of the IEEE, AES, B/PAA, and the Economics Club of Detroit and has been included in "Who's Who in U.S. Writers, Editors and Poets", Marquis "Who's Who in Finance and Industry," and US Registry's "Who's Who in Leading American Executives." He is listed in Bradford's Directory of Marketing Research Agencies and Management Consultants in the United States and the world—the only firm specializing in the metalworking field.